PHONE
WARRIORS

Exposing the Telecom Wars of the 1970s

by

PAUL D. OLSEN

FROM THE TINY ACORN...
GROWS THE MIGHTY OAK

For information, address Acorn Publishing, LLC, 3943 Irvine Blvd. Ste. 218, Irvine, CA 92602
www.acornpublishingllc.com

Cover design by Damonza.com

Author photo by Remy Haynes Photography

Interior design and digital formatting by Debra Cranfield Kennedy

ISBN—978-1-952112-49-2 (hardcover)
ISBN—978-1-952112-48-5 (paperback)
Library of Congress Control Number: 2021903661

Table of Contents

The Myth

"Ma Bell and Apple Pie." These were common, midcentury refrains heard in households and businesses across America, and one frequently used by the Bell System to provide consumers with a feeling of comfort and warmth. Growing up, it was a feeling my family and I shared. But "Ma Bell" was also a term crafted and embraced by AT&T to cloak its monopoly of telephone services in the United States. Make no mistake, Ma Bell was a mean mother—a BEAST!

Much has been written about AT&T, the Bell System history, its break-up in 1984, and the growth of spin-off "Baby Bell" companies afterward. Many magazine and news articles at the time of the break-up were of the doom-and-gloom sort, prophesying the divestiture of the Bell System would result in increased costs to the consumer. However, most books and articles written since the late '60s covered the frustrations experienced by Bell System management and its employees as they encountered competition. Written by former Bell executives, industry insiders, and former FCC personnel, they are great textbooks on organization, rate-making, politics, economics, and business development.

But this isn't one of them.

Phone Warriors lifts the veil from AT&T's backroom deals and activities to stomp out any competition. For the first time, the stories of energetic entrepreneurs are told along with the unsavory situations they encountered. Previous literature has completely missed the energy and chaos in the telecom world during the late '60s, '70s, and '80s. Called "Interconnection," this new era of competition created an industry as vibrant as modern-day Silicon Valley and the internet, a new opportunity that attracted idealistic, energetic entrepreneurs— and opportunists with less-than-ideal scruples.

Contained herein are the untold stories of corporations' and small entrepreneurs' efforts to gain a foothold in AT&T's vast empire. Initially, few recognized the impact a 1968 FCC ruling would have, allowing others to compete in Bell's marketplace for equipment and services. No one expected the nefarious tactics employed by the Bell System to limit competition in what Bell considered their own back-yard. Ironically, the Bell System's own anticompetitive behavior arguably contributed to the U.S. Department of Justice's antitrust suit leading to the eventual break-up of AT&T in early 1982. They may well have been the authors of their own demise.

Paul D. Olsen

Civil Unrest–Rebellion Against Authority

A rebellious movement slowly began to build. Scholars may debate finding a particular moment, event, or date in which the paradigm began. But develop it did, and grew to devastating proportions. In most democracies the pendulum of attitudes, tolerance, morale, equality, popularity, swings from side to side—and back. During the 1940s to the early '60s, acceptance of the political establishment was fairly constant. Political graft, indiscretion, and greed was headlined but often tolerated by the public. The temptation to use political positions for monetary gain was too hard to resist. Transgression seemed part of the business of governing the populace—"Politics as Usual." But attitudes began to change in the early '60s. A strong anti-establishment attitude grew among the younger generation. Part of the fuel for change was the Civil Rights Movement led by Dr. Martin Luther King Jr. and his call for "non-violence" and "resistance through civil disobedience." Officials often met his protest marches and "sit-ins" with violence. Eventually the government made changes including The Civil Rights Acts of 1964 and 1968, and the Voting Rights Act of 1965.

Arguably it was the inequality that the Civil Rights Movement responded to that led to civil discord, but the movement creating the most momentum was the resistance to the Vietnam War led mostly by a large segment of youth who were either still in school or had recently entered the workplace—many subject to the draft. It was a period of open resistance to almost all authority. Young men receiving draft notices requiring them to report to their local draft board would often return them to the Selective Service or burn them. The dissidence against public policies would become violent.

In an FBI report given in July of 1970, before the "President's Committee on Unrest," J. Edgar Hoover stated, "Violence is not Dissent!" The report cited the following: From 1968 to 1969, there were 850 protest demonstrations, 200 government and school building seizures, 4,000 arrests, sixty-one attempts at arson, all resulting in one death and 125 injuries. The following year recorded a significant increase: 1,785 protest demonstrations, 313 building seizures, 7,200 arrests, 246 arson attempts, and fourteen bombing attacks, resulting in eight deaths, 462 injuries, and property damage of $9 million. Referring to the previous two years, the report blamed the acts of violence and disorder commonplace in colleges and secondary schools on extremist forces. Civil disobedience had become resistance of authority and eventually violence and lawlessness. Not only was a segment of youth to blame, but Hoover included "members of the academic community who defend or condone wanton acts of violence and riotous conduct do obvious harm . . . and undermine our Nation's confidence, morale, patriotism and unity." Hoover continued, placing the blame on "arrogant spoiled youth that may not be emotionally mature to enter an institution of higher learning and that are ill prepared to dictate academic standards or policies of a college or university."

The recommendations summarized in the report placed the blame on parents, university administrations, and enforcement agencies for being too lax in their responsibilities and needing to be more stringent. It called for more discipline in the home and universities. Paradoxically, the theme of "antiestablishment" was not lost on the young men and women entering the workforce in the mid '60s and early '70s. This, then, was the backdrop of young professionals in this period who chose careers in the government sector in agencies such as the Department of Justice and the Federal Communications Commission. Upon entering the public sector, their attitude may not have been solely intent on questioning authority or antiestablishment. However, many had open minds to question if current establishment and policies were appropriate for the public and if they should be changed.

• • •

BOOK ONE

The Beginning of an Industry

CHAPTER 1

AT&T History

Shortly after speaking through his transmitting device to Mr. Watson in 1876, Alexander Graham Bell quickly grasped the significance of his invention and commented to his father that people would soon be able to communicate with each other without leaving their home. He envisioned a national telephone system with wires connecting city telephone exchanges everywhere. With a few financial backers, the Bell Telephone Company was formed in 1877. Before Bell transmitted his famous words to Watson, Gardiner Hubbard—a patent attorney from Boston— recognized the potential of Bell's invention and agreed to underwrite the cost of Bell's research for a share in the enterprise. Hubbard quickly filed for a patent on Bell's behalf. A few days after the company was formed, Bell married Hubbard's daughter and left for England. Hubbard was then tasked with finding out how to turn the new Bell Telephone Company into a money-making enterprise. He soon hit on a unique idea: *telephone franchises*. Bell agreed to the concept, and Hubbard began licensing businessmen in cities across the country to sell telephone service using the patented Bell telephone. Hubbard insisted that franchisees only rent, not sell telephones to their customer. This became one of the most strategic decisions ever made in U.S. corporate history.

In 1878, under a license agreement from Bell Telephone, the first telephone exchange in the United Stated began operation in New Haven, Connecticut. Within a few years, local telephone exchanges, referred to as telephone franchises, were constructed in every major city. With the exception of some rural exchanges, most all licensed telephone exchange companies in the contiguous United States were eventually purchased by American Telephone & Telegraph. During the economic depression in the late 1870s, Hubbard relinquished control of the company to a group of Boston investors. Before doing so, he hired Theodore Vail who was superintendent of the U.S. Railway Mail Service. When Vail arrived, Bell Telephone was in the middle of a patent fight with Western Union, the reigning business power of the day who was also offering telephone service using Bell telephones. The suit was settled with Western Union agreeing to sell its telephone business to Bell. Investors soon pushed the price of Bell Telephone shares from a few dollars to $1,000 a share.

Bell Telephone reorganized and acquired the name American Bell. Vail, concerned that the patents would run out in ten years, expanded on Bell's initial vision. Vail saw the future as one national network by tying together all American Bell franchise exchanges with long-distance lines. In 1885, American Bell established a new subsidiary, American Telephone and Telegraph—AT&T. Its charter: build and operate the nation's first long-distance network. In 1899, AT&T acquired the assets of its parent. In the beginning, the quality of long-distance service was lousy. Technical solutions necessary to resolve the problems were not completed until 1915. Regardless, AT&T did not allow independent telephone companies to connect to its network until it was later forced to by the federal government.

Vail believed in the superiority of one national telephone system and that the telephone and its technology should operate most efficiently as a monopoly. Vail's slogan became "One Policy, One System, Universal Service." In the early 1900s, AT&T began acquiring smaller companies, including Western Union—the U.S. telegraph network. AT&T also had investments in other countries. These included the Bell Canada operating company and Northern Electric—a subsidiary of AT&T's Western Electric. AT&T also owned various Caribbean telephone operating companies, owned a majority position in the Japanese manufacturer Nippon Electric Company (NEC), and had a post-World War II reconstruction relationship with the Japanese telephone operating company, NTT. AT&T (the Bell System) would operate as a monopoly for several years. By the mid 1960s, AT&T had become the largest corporation in the world.

Throughout its history, AT&T experienced many challenges to its growth. When a new local telephone exchange was constructed and operated by an independent company, AT&T would refuse to allow the independent operator to connect to the AT&T network—putting pressure on the new company to eventually sell under AT&T terms. In 1913, the federal government—concerned about the Bell System's increasing monopoly over its telephone network ownership under AT&T—filed an antitrust suit. In 1919, AT&T made a deal known as the Kingsbury Commitment. Fearing that the suit might result in a break-up or nationalization of AT&T's network, the company agreed to divest itself of interest in Western Union and allow noncompeting independent telephone companies to interconnect to AT&T's long-distance network. After 1934, the Federal Communication Commission (FCC) assumed regulation of AT&T. Vail had accepted regulation as the price of maintaining and extending the Bell System's telephone monopoly. He did so to satisfy the regulators that AT&T was not making excessive profits from its monopoly.

In the 1940s, the Bell System effectively owned most telephone service in the United States. This included local exchanges, long-distance service, and all customer premise equipment consisting of private branch exchanges (PBXs) and telephones. A key strategy in

AT&T's business plan was to own everything in their network. Subscribers could only pay to use the network and to rent or lease equipment made by AT&T's Western Electric. AT&T even claimed ownership of the wiring inside and outside a subscriber's business or home. If a subscriber wanted a particular type of phone not offered by the local Bell Operating Company, they had to purchase it and provide it to the local Bell Telephone Company at no cost. The local Bell Company might "test" the phone and, if approved, would rent it back to the subscriber for a monthly fee. They might also impose a connection charge.

Again, in 1949, the U.S. Department of Justice filed another antitrust lawsuit against AT&T, alleging that the Bell System and Regional Bell Operating Companies bought all their telephone equipment from Western Electric creating a "captive monopoly." The Justice Department thought AT&T was also using its monopoly position in telecommunications to establish an unfair advantage in related technologies. This resulted in another consent decree in 1956 that limited AT&T to eighty-five percent of the nation's telephone network and allowed AT&T participation in specific government contracts. AT&T also was restricted from holding interests in the telephone operating companies in Canada and the Caribbean. The decree also included the divesting of Western Electric's interest in Canada's Northern Electric that was manufacturing telephone equipment, relying mostly on research and development from AT&T's Bell Labs.

By the early '60s, the "Bell System" referred to all AT&T companies and its four major divisions:

- **Bell Labs**: provided the research and development of products and components it concluded was required for use by AT&T and its operating companies. Bell Labs developed a huge inventory of patents still in use today.

- **Western Electric**: manufactured the equipment and products developed by Bell Labs and distributed them to AT&T and its operating companies.

- **AT&T Long Lines**: provided connections to and between local telephone exchanges and provided long-distance and international calling services.

- **Regional Bell Operating Companies** (RBOCs): provided local exchange calling services and telephone equipment. The RBOCs paid a fixed fraction of their revenues as a license fee to Bell Labs. Each RBOC controlled the rates for equipment and services based upon their perceived value to their customers and as approved by the State Public Utility Commissions.

The Bell System's goals of "Universal Availability and Superior Quality" continued into the 1970s. To achieve these goals at reasonable costs to its subscribers, AT&T claimed to set prices below costs for basic residential phone service and covered those costs with revenues from higher priced long-distance calling and business services and equipment. The company averaged the service costs and apportioned those charges to provide telephone service to the greatest number of subscribers. AT&T developed economic models of "rate averaging" and "separations" to determine service costs and hide subsidies rather than the traditional "cost plus" basis that most businesses employ today. AT&T profits were related to their total invested capital in plant and equipment, and the company was guaranteed a fair or "reasonable" return on investment by the various regulatory agencies.

The Common Carrier Department of the FCC regulated and approved tariffs for AT&T's Long Lines services, considered the most profitable of AT&T's business empire. AT&T's Long Lines argued that their revenue helped subsidize the cost of providing low-cost service to rural areas. The Regional Bell Operating Companies that provided local service and telephone equipment became regulated by State Public Utility Commissions. The RBOCs quickly developed close relationships with the State Public Utility Commissions (PUCs). When the RBOCs felt that a greater rate of return on their invested capital was appropriate, a petition for a rate increase was filed with the State Utility Commission. The "rate averaging" and "separations" types of economic models employed by the RBOCs usually overwhelmed the State PUCs. Rate increase requests, after some negotiation, were usually approved. The net increase averaged between seven to ten percent and as high as fifteen percent. Under the structure of a utility, there was no incentive to reduce or control costs. One PUC regulator described the process:

"They come in requesting a rate increase for more than they need, we both know that. We eventually approve something lower than their request, probably close to what they wanted. That keeps the Bell Operating Company and the public happy."

Most every business determines the price of its products or service to the consumer based upon its material cost, labor, overhead, and profit. The Bell System made a major deviation to this model. As the Bell System maintained ownership of their products, each Regional Bell Company was able to charge for its services and equipment by what the local RBOC thought the perceived value of the service was to its subscribers.

Although the cost of telephone equipment from Western Electric to all RBOCs was virtually the same, depending upon where the subscriber lived or had an office, there was little consistency in what they actually paid. As an example, a basic single line telephone used in a home or behind a PBX in a business rented for $1.00 a month in one city and $2.50 a month in another. A "six button telephone" could rent for $1.75 per month in California and $4.00 a month in Missouri. A multi-button (twelve or eighteen buttons) "call director" telephone in 1971 rented for $4.00 month in California, but for $17.50 a month in Illinois and $22.00 in New York City. The monthly rent for subscriber lines and trunks also varied by state.

In early 1970s, the "Premise Equipment" market consisted of just two basic categories for business telephone systems provided by the Bell System. Small systems with two to twelve business lines/trunks and up to twenty-five or thirty telephones were referred to as Key Telephone Systems (KTS). These smaller systems were supported by a central Key Service Unit or KSU mounted on the subscriber's equipment wall that allowed the user to access a line by simply pressing a button (key) on the telephone to answer a call or dial out. To answer and distribute incoming calls required one or more dedicated people, usually the receptionist. The caller would be placed on "hold" while the receptionist would dial the designated party within the firm over the system's intercom line and inform them which "line" to pick up by depressing the associated key: "Mr. Jones, call on line three." As a business grew, the KTS became less efficient due to inadvertent interruptions and dropped calls, as anyone could access a call in progress, interrupting the conversation or answering a call not intended for them by depressing the wrong button.

Larger telephone systems required a dedicated operator to answer and distribute all incoming calls through a switchboard or private branch exchange. Early manual PBXs, or "cord boards," were still around in the late '60s and '70s and required internal personnel to access the operator and ask for an "outside line." The operator would connect the internal party with a pair of cables to an available trunk port on the PBX. With an automatic PBX, the subscriber simply dialed "9" and would be connected to a trunk line.

With either of the above systems, the cost to rent telephone equipment did not decrease in relation to the number of lines or telephones added. Nor was the monthly cost linear. As a business expanded and required more business lines/trunks and telephones, the cost increased exponentially. On a per-station or per-user basis, the monthly cost from the telephone operating company for equipment rental could triple.

Within AT&T, operating as a monopoly with a rate of return periodically granted by state and federal regulatory agencies, there was little incentive for improvement and innovation. Nor did AT&T have any structure to receive input from subscriber demands. Market influences were alien to AT&T's way of business. Bell Labs determined what the market wanted and developed products that were then manufactured by Western Electric. AT&T determined what was best for their subscribers. There was never a thought that subscribers were customers who might want more features or other choices.

During the late '60s, subscribers began to experience a degradation of service in metropolitan areas. Most severe were the outages of central exchanges in New York, the capital of commerce and financing for AT&T. Corporations including investment banks that supplied funding for AT&T found that it was taking longer and longer to repair service outages. This came at a time when technology was expanding and the benefits of computing power were being discovered and appreciated. In 1968, when the FCC decided that others could compete with AT&T for equipment and services, the Bell System's premise equipment market was ripe for competition.

The way Bell Operating Companies protected its monopoly of telephone equipment was exemplified by a story told to me by a San Francisco attorney while I was selling dictating systems in San Francisco in 1964—before entering the telecommunications industry. The elderly attorney worked in an old building in a two-man office in the Marina District. His partner had retired years earlier. One light on his telephone no longer lit up or flashed to indicate when a call came in. His secretary, Margaret, requested a service call to Pacific Telephone that had been scheduled later in the morning.

Margaret announced the telephone repairman's arrival shortly after 11:00 a.m. The attorney described the problem. Pointing at the telephone on the attorney's credenza, the repairman asked: "What is that on your telephone handset?"

The attorney turned to see what the repairman was talking about. Gesturing toward the telephone with a suction cup stuck on the handset he replied, "It's an induction device that I use to record calls with my clients."

"That's Bell System property, and you cannot attach anything to it!" came the terse reply.

"Well, I've been doing it for years and was never told I couldn't record private conversations."

"Sir, you will need to remove it, or I will need to remove the telephone now. Nothing is to be connected to Bell System equipment."

The attorney easily snapped off the suction cup—it was always falling off anyway. The repairman quickly repaired the problem with the light on line two and left before noon. As the outer door closed, the attorney turned in his chair, licked the suction cup, and re-affixed it to the handset. He was back doing business as usual.

The attorney, like most of AT&T's subscribers, was not aware of FCC Tariff No. 132 (modified in 1957) which read: "No equipment, apparatus, circuit or device not furnished by the (AT&T) Telephone Company shall be attached to or connect with the facilities furnished by the Telephone Company, whether physically, by induction or otherwise. In case any such unauthorized attachment or connection is made, the Telephone Company shall have the right to remove or disconnect the same; or to suspend the service during the continuance of said attachment or connection; or to terminate the service."

AT&T had lobbied the FCC well, and the FCC staff may not have considered how this petty restriction would inhibit the use of devices beneficial to subscribers while not injurious to AT&T's network.

CHAPTER 2

Thomas F. Carter—Part I: Challenge and Frustration

The story of the Texan, many thought a cowboy, who changed an industry.

Dallas, Texas, Late '50s

Was he a cowboy, cattle rancher, or electronics tinkerer? Tom Carter, an extrovert to many of his friends, was of average height about five foot eight inches, stocky build, broad shoulders with a full head of premature white hair. Known for his aggressiveness, he used his training in the Air Force to start an electronics firm in Dallas in his early twenties. Several articles have been written portraying Carter as a rancher—riding his horse on the range, perhaps having more than one ranch before developing his Carterfone. I had even heard the story of his missing an important telephone call while out on his range that contributed to his development of an instrument that would help him with his livestock business. "Not True!" claim family members. "He was an urban businessman who was a member of just about any organization such as the Lions Club, Rotary, or a businessman's breakfast club where he could network and stimulate his business."

Carter fully understood the need for improved communications in rural areas in Texas. He had talked to ranchers and oil field men who needed to communicate with their customers and suppliers while tending their livestock or repairing pipelines. Carter's electronics company in Dallas sold two-way radios and paging services to these industries. Through the two-way radios, dispatchers could communicate with those in the field miles away from the office or a landline. At the height of Carter's Dallas electronics company, he had about 100 people working, selling, and installing radio towers and mobile radios. He had installed 300-foot towers in seventeen western states—each covering a fifty-mile radius. Carter would fly his own twin engine Beechcraft Baron for his business and was considered by his peers as "well-off," an affable, but a tough and stubborn businessman. "He couldn't just let go of something that irked him—and he could be arrogant and mean if he felt you crossed him." Carter recognized a greater need for those in remote places to be able to connect to the telephone network via a private two-way radio.

He claimed to have been following the "Hush-A-Phone" case filed with the FCC in 1948 for almost seven years. For twenty-five years, Hush-A-Phone had been making and selling a soft rubber cup device that could fit over the mouthpiece of a handset that allowed the user to speak into the telephone and not be heard by others nearby, ensuring privacy. Citing the Communications Act of 1934, AT&T objected to the device as a "foreign attachment" and threatened discontinuance of service to subscribers who used it. The 1934 Act forbade the attachment to the telephone of any device not furnished by AT&T. AT&T argued that the device could damage the network, resulting in a general deterioration of the quality of telephone service. Initially, the FCC upheld the ruling. Hush-A-Phone appealed. AT&T failed to show how such damage could be created, and in 1956, the D.C. Circuit Court of Appeals decided in favor of Hush-A-Phone and prohibited further interference by AT&T. The court found that AT&T's prohibition of the device was not "just, fair, and reasonable as required under the Communications Act of 1934." Further, the court found the FCC tariffs regarding the Hush-A-Phone an "unwarranted interference with the telephone subscriber's right reasonably to use his telephone in ways which are privately beneficial without being publicly detrimental."

Carter claimed in a September 1984 "Communications News" article that he began working on the Carterfone as early as 1949, back in the days of the vacuum tube. That would have put Carter at age twenty-five. "That's bullshit!" states Wallace Hammond, Carter's son-in-law, adding: "Tom Carter told me the story several times. He was an electrical genius. A truck driver customer of Tom's two-way radio systems told him how he wanted to be able to talk to someone on the phone while in his truck. Tom thought about it, and it only took him just a couple of days to develop it—he might have done it overnight."

As an aside to his regular radio business, Carter started working on his Carterfone device that could provide a connection between a telephone handset and a two-way radio. It was an induction device that required no direct wiring to the telephone company's handset. Carter thought the Hush-A-Phone decision was the door opener for his new product. He had a ready market. "We just showed up at oil company shows and started selling," Carter said. The device required a person (usually a "dispatcher") at the two-way radio base station who, upon receiving a call on the base station radio, had to manually dial up a number on the telephone, place the telephone handset into the Carterfone acoustic coupler, and press a button to initiate the conversation connecting the base station to the telephone—through induction, just air. It would also work in the reverse mode. Automatic systems came much later.

Carter had a hit on his hands, and selling Carterfone was easy until AT&T came calling. Carter subsequently found out that the

FCC had merely shelved the court-ordered Hush-A-Phone tariffs. They never went into effect. AT&T (again) claimed the Carterfone device a foreign attachment. As Carter stated, "The phone company began harassing my customers—threatening to cut off their phone service unless they quit using the Carterfone." Carter tried logic on AT&T, as he knew his device was not harmful to their network. Logic didn't work. He visited the FCC, where he learned it would probably take an antitrust case to get AT&T off his back. Then he went to AT&T headquarters with a proposal: Carterfone customers would agree to pay AT&T $1.00 per month to use his device. In turn, he would agree to not initiate an antitrust lawsuit against them. The offer was refused. "Had they taken the offer," Carter claimed, "they wouldn't have competition today!"

CHAPTER 3

Telecommunication Consulting

Large and small businesses, hospitals, colleges, and non-profit and government agencies were beginning to realize the extent of escalating costs of long-distance calls and the rent paid for their telephone equipment. Bell System representatives were quite efficient at recommending more trunks, WATS (Wide Area Telephone Service), Tie Lines, T-1 spans, foreign exchange lines, 800 numbers, and additional phone equipment to their subscribers. Reduction of a subscriber's telephone expenses was rarely the objective of the Bell System and its marketers. Seldom did a subscriber reduce their telephone operating expenses through the Bell System's recommended changes or additions. Simply stated, the phone company never had any incentive to reduce a subscriber's telephone expenses.

An industry grew consisting of telecommunication specialists who could, for a fee, render advice on how to manage and control telephone costs. Some large companies recruited full-time telecom managers from the local Bell Operating Company. The consultants focused on long-distance call patterns to justify 800 numbers, foreign exchange lines, handling the details of a move, recommending specific Western Electric telephone equipment, auditing billing for equipment rentals, restricting phones, or enhancing call processing to accommodate increased service to the customers of the business or agency. Some consulting firms specialized in redesigning the system and augmenting it with an ancillary paging or intercom system that did not connect to the current phone system. Other telecommunication consultants who focused on over-billing and billing errors based their fee on a portion of any identified billing errors refunded back to their client by the operating company.

The Bell Operating Companies charged a monthly lump sum to all subscribers from an inventory listing all equipment, lines, and trunks. Calls made by the subscriber were billed separately. The first step with most consultants was to reconcile the subscriber's monthly bill to the operating company's list of inventory for the subscriber. Consultants identified billing errors with ninety percent of their clients. It had always been the responsibility of the subscriber to inform the telephone company in writing of the equipment and services it no longer required. Otherwise, the telephone company would continue billing for the service even though it was not being used. Common billing errors were discovered in:

- Business lines or trunks defective or no longer used.
- "Night connections" to answering services that had been discontinued or replaced with voice mail.
- "Off premise" extensions, foreign exchange lines, and "tie-lines" to facilities that had been moved or consolidated.
- Data circuits that had become obsolete.
- Lines and equipment to departments that had been moved or removed.

During an installation of a private telephone system for the Baptist Memorial Hospital in Kansas City, eight "dead" incoming trunks were identified. The hospital estimated they had been paying for the unusable trunks for eight to ten years. Southwestern Bell technicians had simply wired around the dead trunks as more trunks were added. In another instance, a manufacturer in St. Louis had several boxes of telephones removed after they installed a paging system. The owner claimed the Bell technician "temporarily" placed them in a storeroom for pick up later. They were never picked up. The Bell Operating Company continued to bill the subscriber for the unused telephones every month—for years. These were just two of thousands of examples of Bell System subscribers being charged for services they were not receiving.

Telephone consultants were considered an irritation by most Bell System marketing and technical personnel. To perform their work effectively, the consultants were required to obtain a "Letter of Agency" from their client addressed to the local Bell Operating Company. The Letter of Agency stated that the consultant could act on their behalf in all matters relating to the subscriber's telephone system and requested cooperation in complying with the consultant's requests. The consultant would request an inventory record of the equipment billed to the subscriber, and then conduct a physical reconciliation of the list. Once a billing error was discovered, the consultant would notify the Bell Operating Company of the error in writing. At their leisure, Bell would dispatch a technician (sometimes accompanied by a Marketing Representative) to the customer's office to validate the discrepancy. An on-site meeting between the consultant and a Bell System representative was necessary to validate the billing discrepancy.

Meeting with a Bell telephone representative at the subscriber's site at a specific date and time to obtain a refund required a major effort. If the consultant did not meet the technician on the subscriber's premise, the technician's validation report would often state, "no billing discrepancy was found." When billing errors were validated, the consultants would request a refund check made out to the subscriber and delivered to the consultant who would hand deliver it to the client with an invoice for his services. That worked only for a short while. The Bell Operating Companies soon began filing petitions with the State Utility Commissions (PUCs) to limit their liability. Many State PUCs allowed the Bell Companies to limit their exposure to overbilling and billing errors to only three years. When refunds were acknowledged, in spite of the consultants' instructions, the refund often showed up as a credit on the client/subscriber's phone bill several months after the billing errors were confirmed. There was never a cover letter or anything in writing provided to the client or consultant admitting to the billing error or explaining that the credit would be provided in a future bill. The credit, when issued, would show up on the client's monthly statement two to three months after verification of the error(s). One can imagine the difficulty the consultant had in collecting a portion of the refund without documentation.

Not all consultants were ethical. After the FCC permitted subscribers the opportunity of choosing telephone products from different manufacturers other than Western Electric's through the Bell Operating Companies, some consultants took advantage of the situation. Some directly asked for a commission if they recommended a specific product from an authorized distributor, rationalizing that the consultant was performing the function of the distributor's salesperson. Some telephone distributors eager to win a bid occasionally offered bribes. Wining, dining, and other inducements offered to consultants who controlled large potential sales were not uncommon. Exposure of the practice came to public light with the investigation of rigged telecommunication bids to government agencies in the late 1970s and '80s, some of which are covered in subsequent chapters.

CHAPTER 4

The New Frontier

One man takes a risk, leaving a secure position to start a company in
a new industry he knows nothing about. And I decide to join him—
knowing nothing and finding out what I didn't know.

Cabo San Lucas, Mexico, 1966

He dropped himself on his blanket after a long swim in the warm Mexican Gulf and began reviewing his goals in life. After his service in the Coast Guard, Craig Dorsey had begun a successful career as an office products salesman for IBM in San Francisco selling typewriters, dictation systems, and supplies. He was a hard worker and ambitious. While IBM held several classes in salesmanship, he improved his skills by completing a Dale Carnegie Sales Course in the evenings. He returned as an unpaid assistant instructor in subsequent evening classes, as it continued to improve his sales ability. Dorsey had been with IBM for three years and, after becoming "Salesman of the Month" for the district, had decided to take a vacation in Mexico. Dorsey wondered how far his career with IBM would take him and if he should make a change.

One customer in Dorsey's sales territory, Phone Consultants, Inc., was small but growing. The sales staff, management, and owners were all about Dorsey's age. They were sharp, excited, and creating a working environment similar to that he experienced at IBM—which he loved. The two owners of the firm, Tom Hingle and Bob Rengo, suggested Dorsey join the company. The two of them had just bought new Chrysler convertibles—a sure sign of their company's recent success. The business was appealing, and the place seemed to be on a roll.

While in Mexico, Dorsey made his decision and flew back early. He informed management of his decision to resign IBM and was told he would be welcomed back should the new job not work out. He agreed to work another month and submitted his resignation. After his last day at IBM, he drove over to the Phone Consultants building in anticipation of interviewing for a sales position. The new Chryslers were gone. The office door was chained closed, and no one was inside. It didn't appear that the business would soon reopen. Dorsey had just given up his career with IBM for a business that appeared defunct.

One of the Phone Consultant salesmen, Jerry Williams, was just coming up the outside stairs. Dorsey didn't know anything about the telephone business, or consulting for that matter, but Williams had some experience in the industry. They moved from the parking lot to a coffee shop to discuss the situation. After some deliberation, the two of them decided to approach the owners to see if they could buy the company. Hingle and Rengo owed about $100,000 to creditors and wanted an equal amount—or $200,000—for the company. Dorsey and Williams thought the offer excessive and decided to form a partnership together. The two called their newly-formed company Communications Consultants, Inc. and rented space in an old carriage house on Filbert Street in San Francisco.

If you worked for IBM and did a reasonably good job, you had it made. I worked for IBM competitors selling office products in the financial district of the city. The financial district in San Francisco was a tough place to compete against IBM. Having been a branch sales manager for a copier company the last three years in three cities, I, too, felt the need for a change.

BECOMING A TELECOMMUNICATION CONSULTANT

The stoplight on Howard Street had just turned red. As I sat in my car at the light, a new Oldsmobile convertible came up next to me—top down. I recognized the driver, a competing office equipment salesman I saw often on the street and whose sales ability was regarded as adequate, but not great.

"Irv, how are you doing?" I hollered over to him.

"Great!" he replied.

Wondering how he could afford the new Olds, I asked, "What are you doing now?"

"I'm a telephone consultant!" shouted Irv.

"Pull over to the curb. I want to talk to you," I replied.

Irv told me he'd joined a telecommunication consulting firm a few months before, Communication Consultants, Inc. (CCI), owned by Craig Dorsey and his partner. I knew Dorsey from the evening Dale Carnegie classes where I often worked with him as an assistant instructor. After a lesson, Dorsey and I would divide the class into two sections, each of us "role-playing" what had been taught that evening. From my curbside conversation with Irv, the telephone consulting business sounded intriguing and appeared to offer a greater challenge than selling typewriters and copiers. I recalled an earlier conversation I'd had with a college friend, Don Mall, whose work sounded similar. I made an arrangement to meet him for lunch a few days later. Don worked for a competitor to Dorsey's new company, E.D. Jones, in Oakland. I peppered him with questions about the business. It sounded much like what Irv had described.

Deciding I needed more technical information before seeking work in the industry, the next day I walked into the Pacific Bell office in San Francisco and naively announced that I wanted some brochures on their office phone systems and pricing for the telephone equipment and service they offered to business customers. From the look on the receptionist's face, it was a most unusual request. It was some time before two representatives appeared before me. I repeated the request. Young, trim, good looking in their tailored suits, they asked in a not too friendly tone, "Why?"

Images of being escorted to an office or conference room with the two Bell representatives sharing the information I desired quickly vaporized. The three of us were standing in a small, glassed vestibule next to the receptionist area.

"I'm doing some research on your industry," came my weak reply.

"Where is your place of business?" was their next question.

I had to acknowledge that I didn't have one and asked if they had some basic pricing information they could provide. The taller of the two turned and pointed through the glass to several shelves of thick black binders.

"Those are just some of the tariff books that cover our products and services." Turning back to me, they added, "When you have an office and an address, we will be happy to send a marketing representative to your location to better understands your needs. At that time, Bell System products will be recommended to you. Otherwise, we really can't help you. Thanks for your interest."

Both turned in unison and walked out of the vestibule. My learning curve on Bell telephone system equipment had just taken a sharp upward turn.

INITIAL SALES

David Collins looked up from his desk toward the direction of the loud noise. Collins owned a growing motorcycle dealership in the East Bay and was working at his desk late that night. From his office ceiling protruded a leg which hadn't been there a moment ago. Dust from the broken ceiling tile had fallen on his desk.

"Who's there?" he demanded.

The dealership had outgrown its telephone system and had contracted with a telephone consulting company. Earlier that day, Dorsey and Williams, representing their new consulting company, CCI, had made a presentation that included a demonstration of a "hands-free" intercom and paging system made by L.M. Ericsson of Sweden. The consultants' recommendation would be more efficient for distributing calls and internal communications. It was complementary to the phone system and competitive in price to the system proposed by Pacific Bell. Mr. Collins signed their sales contract.

"It's us, Mr. Collins, Craig and Jerry," came the reply from above. "We're putting in the cabling for your new communication system." Collins hadn't noticed the two returning later in the afternoon and carrying material to begin the installation process into the equipment room. References were important to CCI's success, so Dorsey and Williams had decided to complete the installation of any new sales as quickly as possible. As he walked to his car, Collins began to question his earlier decision to contract with the small unknown company and began thinking of questions he should have asked during their presentation.

Shortly thereafter, the duo made another presentation to a prestigious San Francisco law firm. The contract was signed late in the afternoon. Everyone wore a suit and tie in the city during the '60s—probably still do. The ink still wet on the contract, Dorsey and Williams made a hasty retreat to the car for a change of clothes as they intended to start the installation that night. It was prudent to "spike" a job, or commence installation before the customer had second thoughts or competition surfaced. In the midst of changing clothes in the law firm's restroom, one of the partners who signed the contract walked in giving them a quizzical look.

Sheepishly, Dorsey explained, "We decided to start the installation tonight."

"You guys really move quickly—we could never get the telephone company to act that fast," declared the attorney.

They continued working into the night. The company was so new that they had not yet hired any trained technicians, and the Ericsson intercom system was new to them. Regardless, Dorsey and Williams decided to install it themselves and began reading the manufacturer's installation manual in the equipment room.

In the early morning hours, the wiring had been completed and the central control unit mounted. Next, they needed to plug in and test the Ericsson intercom stations. As every location had to be tested, Dorsey began the process by calling Williams who remained by the central unit. "Hear you fine," was Williams' usual reply. Dorsey then moved to the next station. The sun was starting to rise. Most of the intercom stations had been tested. They needed to finish the job, go home, and clean up so they could return and instruct the lawyers and their staff on how to use the system properly. Dorsey was testing the last two stations. From the station, he dialed Williams back at the control unit. No answer. Dorsey checked the wiring. He then used a new station. Again, no answer. He moved to the next location and repeated the test. No answer from this station either. Dorsey retraced his steps. He found that the last intercom station that tested well was not working. Something had gone wrong, and the employees would be arriving in a few hours. Frustrated, he changed the location of stations, reread the instructions, and again checked the station wiring. All Dorsey's work appeared to have been completed correctly. Now confused and tired, Dorsey walked back to the equipment room to tell Williams of the failures. On a stool by the central unit, Williams was sound asleep, head against the wall. He hadn't heard any of the last tests. Regardless, the law firm became a great reference. CCI was off to a good start.

TRAINING

A loud bang ceased all conversation in the office. Dorsey was making a presentation to a group of eight people in the conference room of a company that earlier had engaged CCI to assist them with their telephone system needs. It was at a crucial point in the presentation. All was quiet and everyone was looking at me—with a surprised look on their faces. My mouth was agape, legs dangling over the edge of my chair. Williams wasn't the only one who could fall asleep on the job. The look on Dorsey's face was not one of surprise. I had been leaning my chair back when I fell asleep and crashed into the wall.

Dorsey had hired me a few weeks earlier and was training me in the mornings . . . very early in the mornings. "I'll teach you the business, Paul." Dorsey said, "but not during working hours." It was an adjustment. San Francisco in the late '60s was a great place to work. Dorsey, Williams, and the rest of the sales staff except one were all single guys. We worked hard and played hard. At the end of the day—about 6:30 p.m.—we would retire to a bar on Fillmore or Union Street. Not an uncommon practice in the '60s. There were several bars and restaurants near our carriage house office on Filbert. We made sales calls and appointments during the day and worked on proposals in the evening. This was the work ethic imposed by Dorsey who had an amazing constitution and drive. Before he married, regardless of how late in the evening he might have been out or how bad he felt the next day, he always showed up early in the office. Those who adopted his work ethic became successful. We had to learn the capabilities and limitations of Western Electric's business telephone systems—more thoroughly than the Pacific Bell's marketing people knew. Competing with the telephone company became fun. The business grew.

I was still finding out what I needed to know about the business when a lanky individual showed up at our office. His name was Dave Martin, and he worked for National Communications Planning Services (NCPS) in San Jose. The company was owned by John Barbour, Bud Toly, and Darrel Nelsen. NCPS sold systems much the same way as CCI and distributed similar communication systems. Martin was tall and thin with gray hair and an authoritative voice. He had secured contracts with three Chrysler and Dodge dealerships in the San Francisco area. Barbour reached an agreement with Dorsey to install and maintain the systems Martin had sold. Barbour had already paid Martin a commission advance of $10,000.

Waiting to meet with Dorsey who wasn't back from an appointment, Martin introduced himself and suggested the two of us go to lunch. He was a bit of a legend as a salesman. Knowing I was new to the industry, he offered his experience and asked if I'd like some help on deals I was working on. We reviewed two prospects I was working with—one was another car dealer in San Jose. Appreciating his input, I picked up the lunch tab and thanked him for his suggestions. The next day I found Martin had called on both accounts, attempting to dislodge my efforts and insert himself in the sales process. Nice guy. "Just what kind of people are in this business?" I wondered.

The answer came later in the week. Our lead installer, Chuck Booth, was pulling cable in one of the dealerships Martin had sold. Watching Booth strenuously pull cable through a conduit, the dealer remarked, "This is sure costing someone a lot of money." Booth reported the comment back to Dorsey who called the dealer to confirm his signing of the contract and terms of the lease agreement. He discovered Martin had forged each of the three dealer's names on the contracts—they were not valid. Incensed, Dorsey called Barbour who, in turn, angrily confronted Martin. Martin confessed. When Barbour demanded return of the commission advance, Martin claimed he no longer had the money. Barbour accepted the title to Martin's home as security until the $10,000 was repaid, but Martin had forged his wife's name on the title assignment. A pattern was beginning to appear.

During the first months of training, we would usually request one of the more experienced in our office to accompany us on the closing presentation. Noticing that we had the most success when Dorsey accompanied us, I tracked his personal closing presentations for the next twelve months. Dorsey closed 100 percent of all presentations during the year regardless of whom he was paired with. Using techniques learned in Dale Carnegie, I set an income objective for the following year and charted a formula of steps to achieve it. At the end of the year, at age twenty-seven, I had exceeded my goal. As a single guy working in the city, I did what every young man should do with his extra income. I invested it in my first Maserati.

CHAPTER 5

The FCC Decision—Beginning of the Interconnect Market

Tom Carter fought his antitrust suit with AT&T for the better part of ten years. AT&T threatened his customers and disconnected service to some. A large number of orders for Carterfones had been cancelled by customers and dealers due to the conduct of AT&T and General Telephone, an independent telephone company. Carter's Dallas business began to suffer. AT&T argued that "primary jurisdiction" was vested in the FCC who had the authority to resolve all questions relating to the justness, reasonableness, validity, and effect of the tariff and practices that were complained of. The Circuit Court of Appeals concluded that a decision in the case rested with the FCC but stated, "It is difficult to believe that the FCC, from the communication-utility-point of view, could uphold as just and reasonable . . . a tariff proved to have been brought into being by a concerted action whose aim was monopolistic, not the maintenance or improvement of telephone service."

Upon review of Tariff No. 132, the FCC concluded it had erred in permitting the tariff and rendered a favorable decision for Carter in 1968. Soon thereafter, AT&T successfully lobbied the FCC indicating they would agree to allow foreign devices, but protection of the network was necessary. It was not until 1969, however, that the Bell System filed new tariffs to allow the interconnection of privately owned equipment to the AT&T network through "protective connecting arrangements" designed to protect their network. As a result of the 1968 FCC *Carterfone* decision, AT&T's door was now open to competition. However, the Bell System still felt they owned and could control the door—to the point of nailing it shut. AT&T's other shoe was about to drop.

AT&T'S FIRST BARRIER TO COMPETITION

AT&T's other shoe fell soon after the FCC *Carterfone* decision. AT&T had fought long and hard to keep competition from what they felt was their market and their sole responsibility of providing telephone service and equipment—and theirs alone. They would not give up the fight for several years, even if it might mean fighting dirty. After the FCC's favorable decision for *Carterfone*, AT&T successfully made the argument that if interconnection to their network was to be allowed, then that network must be protected. To the FCC, this sounded logical. Other methods of standardization and registration were not given any in-depth consideration as AT&T lobbied hard against those alternatives. The FCC may have concluded these alternatives would add an unnecessary bureaucratic process that would require time and resources to establish, regardless of the fact that most foreign "private" telephone equipment had been designed around Bell System specifications.

In the first part of the twentieth century, AT&T had interests in Japanese- and Canadian-operating telephone companies. Telephone equipment made by NEC and Northern Tel was influenced by Western Electric specifications. The first private telephone systems installed in the U.S. during 1969 were fully compatible with AT&T's network, and not harmful to it, and yet "protective couplers" or "PCAs" were required to interconnect them. It is ironic that the same PBXs sold as private telephone systems were, in fact, installed in non-Bell Independent Telephone companies without the need for PCAs or protective couplers. What was not covered in the FCC tariff was the monthly rate the Bell Operating Companies could charge the subscriber for each protective device. When a private PBX was to be installed for a customer, AT&T required a protective coupler for each trunk provided by the Bell Operating Company. Depending upon the Bell Operating Company, these could range from $5.50 to $19.00 per month per trunk. They also imposed an installation charge. These couplers were not necessary for any AT&T provided equipment. This additional charge became a penalty for the subscriber choosing a non-AT&T provided phone system. A business subscriber with ten trunks or business lines could easily pay an additional $150 per month for the privilege of owning their own telephone system.

CHAPTER 6

Arcata Communications, Inc.– The First National Interconnect Company

When the FCC decision opened up the market, few national companies were positioned or prepared to take advantage of the opportunity. The following describes one company that held no national presence but moved quickly to establish itself.

Arcata National, a publicly traded timber company headquartered in Palo Alto, relied on its redwood forests for income and profit. Dorsey made a presentation to Arcata for their new offices in Palo Alto that included a reconfigured PacBell telephone system and incorporated a hands-free Ericsson intercom system. Arcata executives concluded CCI's proposed system was more efficient and cost effective than the system proposed by Pacific Bell. Arcata and the Federal government had recently settled a long-standing lawsuit in which the government decided to declare a large portion of Arcata's forests as a National Forest. One Federal caveat required funds from the settlement be used to invest in technology-driven industries. Arcata became interested in the market potential of Dorsey's small company, CCI. Shortly after the installation of their new telecommunication system and the announcement of the 1968 *Carterfone* decision, Arcata bought CCI.

I didn't see it coming, but the following year I would also make a change. In the mid-'60s, I had moved from the city down to San Mateo, near the San Francisco International Airport, where I wrongly assumed many flight attendants (stewardesses, as they were called then) lived. A friend and skiing buddy, "Corky," Keesling suggested I buy a house in the city or in Marin County. He was willing to pay rent to offset my mortgage payment. While looking around Sausalito, he spotted an apartment complex high on a hill in the north part of town, then called the Ardilla (now called Sausalito Towers). He convinced me that we should at least check it out. Most were two-bedroom, two-bath units. The property manager took us to a unit on the top floor. It had stunning views of Tiburon, Angel Island, and the bay, which sparkled in the afternoon sunlight. I was only mildly interested, as it didn't meet my objective of building equity in a home.

While admiring the vast view from the apartment deck, the property manager said, "There are four units of married couples, two units are rented by bachelors in med school, and the rest of the ninety units are filled with flight attendants—averaging four per unit."

"We'll take it!" we both said in unison. She hadn't yet told us what the rent was.

It was almost like living in a sorority. We didn't have to travel far for dates. We lived happily in the Ardilla Apartments for three and a half years.

After work, several of us who were single would meet for a drink at Perry's or The Cooperage in the Cow Hollow district of San Francisco. It was at these venues we would plan for the coming ski season. If we had enough participants—and we always did—we would rent a cabin in Lake Tahoe for the season. The more who committed, the nicer and larger the place we could afford. There was only one rule: Whoever got there first at the end of the week had first choice of the bedrooms. Dorsey's partner, Jerry Williams, Corky, and I left work early Friday and dropped our bags, claiming our space. Shortly thereafter, a casual friend and salesman for Xerox, Howard Smith, arrived with a stunning young lady, Susie, who had just been accepted by American Airlines as a flight attendant and was scheduled to start training in a few months. I didn't need another flight attendant in my life, particularly one living with her parents in San Francisco. This one, however, was too cute to pass up. So, while her date scrounged for sleeping quarters, I learned Susie was working as a receptionist for Utah Construction and Mining Company in the city until she had to leave for stewardess training. The following Monday, I showed up at her work and asked her out for a drink when she got off.

She was shy, kind of quiet, but I found her very easy to be with and loved her smile. We had several dates before she left for training and before the relationship had time to blossom. After training in Dallas, she was assigned to the New York base and shared an apartment with four other attendants. For several months, she continued to occupy my mind. We wrote each other a few times, but as she pointed out much later, my letters were more about business than they were about her. I finally decided to call her mom in Pacific Heights and was told she was flying home that evening for a quick visit. I made arrangements to meet her the following afternoon. She was wearing a gray pin-striped one-piece outfit that complemented her figure. We drove to a park in the Marina District and walked for several hours. She wasn't shy anymore; she was effervescent. Susie did most of the talking while I began to feel queasy, thinking I might be ill at any moment. The colors in the Marina seemed brighter than usual. I made dinner reservations at Rico's, a local's Italian restaurant in Sausalito. The food at Rico's was always good. We ordered a nice bottle of wine. In my late twenties and earning a decent income, thoughts of any serious relationship were far removed. No matter—I couldn't eat. Lightning had struck, and I was a goner. I proposed to her outside the restaurant that night.

She was in a relationship in New York and wouldn't commit. I pressed and asked her to put in for a transfer to the San Francisco base—which is very senior, with lots of flight attendant transfer requests, while she was very junior. Miracles do happen. To our surprise, her request was approved a week later. No other flight attendants at the time had their name on a list requesting a transfer to SFO when the most senior attendant quit. We were married in the fall.

Susan Olsen (Susie)

Shortly after our dinner at Rico's, an exterior wall collapsed, and the place was razed. Just before the wedding, I surprised Susie by buying us a home in Strawberry Point in Mill Valley. First lesson in marriage: Don't ever do that without first consulting her.

Meanwhile, CCI went through its own transition, renamed as Arcata Communications, a subsidiary of Arcata National. Now that it was legal to connect non-Bell (non-Western Electric) telephone systems to the AT&T network, Dorsey's new charter was to build the new company as quickly as possible into a national organization that would sell telephone equipment in competition with the Bell System. He and an officer from Arcata National, Don Thompson, began by acquiring communication firms Dorsey knew about in other cities that were currently competing with AT&T's local Bell Operating Companies or that had the technical know-how to be introduced to this new market. Arcata bought Phone Consultants of Portland from Bob Rengo, the same principal who had once owned Phone Consultants in San Francisco where Dorsey intended to work after leaving IBM. Arcata also bought Rengo's ex-partner, Tom Hingle's company, Phone Consultants of Miami. He had moved to Miami after their business in San Francisco closed.

Before Arcata entered the business, John Barbour of NCPS had sold a communication system to a customer in San Jose that wanted a similar system installed in their Chicago branch. When Barbour and his partner, Bud Toly, flew to Chicago to make the arrangement, they learned the Illinois Bell tariffs for small business key telephones in the Chicago area were tremendously more expensive than in California ($2.00 vs. $10.50 per month for six-button telephones and $4.00 vs. $17.50 per month for twelve-button phones). They thought they'd found a pot of gold. Barbour exclaimed, "We're moving."

Barbour and Toly had made arrangements with Dorsey at CCI to take over their customer base in San Jose and complete installations in their backlog as they moved the business. Arcata later purchased NCPS in Chicago and, in the next few months, also bought small companies in Seattle, Los Angeles, Dallas, Houston, Boston, Atlanta, and New York. Almost twenty locations were acquired in less than eighteen months. I was promoted as Regional VP with profit responsibility for the Northwest Region, which included San Francisco, Oakland, San Jose, Sacramento, and the recently acquired companies in Portland and Seattle. Within a few months, I opened an additional office in Denver, promoting and moving a successful San Francisco salesman as the Branch Manager.

LOTS OF FISH, NO POLES–
THE SEARCH FOR EQUIPMENT

As we started building a national presence, there was a problem. We did not have direct access to telephone products from any of the major U.S. manufacturers: AT&T's Western Electric, GTE's Automatic Electric, and Stromberg-Carlson. We contacted each of them, but the answer was always the same: "We only sell to our parent company and telephone utilities." At least that was true in the beginning. In the early '70s, Japanese manufacturers such as NEC (Nippon Electric), Hitachi, and OKI were struggling to gain a presence in the U.S. NEC had sold a few systems through some small dealers. Our first experience with Japanese PBXs was that the tone of user voices was thin and higher in pitch. We wanted a manufacturer that could give us support and provide on-site technicians and training.

L.M. Ericsson of Sweden had been selling their crossbar PBXs to United Telephone Company. As CCI had been a distributor of Ericsson intercom systems, it might be possible to buy from them. Ericsson agreed to sell to us, but they did not have any presence in the U.S. where product was stored or shipped. They'd been selling to GTE, though, so an arrangement was made where Arcata could buy Ericsson PBXs through GTE. Other enterprising individuals were trying to import small "key" type telephone products from Japan. Arcata Communication held its first national sales conference at Glen Cove, New York in the spring of 1970. All regional VPs and sales managers were invited to attend, and Ericsson was invited to make a presentation of their PBX line. During the conference, we committed to purchase PBXs from Ericsson for all branches. We still needed a smaller KTS system to distribute.

The second night of the conference, Dorsey confided to Butch Shaftsky, Arcata sales trainer, and me that we were to have an off-site meeting with Tom Kelly who had recently joined MEC as its CEO. MEC had begun importing Japanese key system telephone products from Meisei Electric. We were to keep the meeting confidential, as we didn't want to jeopardize our relationship with Ericsson or involve too many Arcata VPs who would want to voice their opinion. That would come later.

Kelly had equipped a van with various Japanese telephones that he could demonstrate to a prospective distributor sitting in an executive office chair in the back of the van. It was late at night. We had eaten well and consumed a fair amount of alcohol. Kelly arrived on time. Dorsey suggested we move to a more discreet location. Shaftsky and I crowded into the passenger seat, and Dorsey chose the executive chair in the back—which was on rollers and not bolted down. Kelly, knowing he had limited time to demonstrate his products, drove rapidly in the dark night taking some Long Island corners a little too quickly. Dorsey rolled in the executive chair, crashing into the sides of the van like a ball in a pinball machine, tipping over and trying to regain composure—not an easy task in a rolling chair, in a moving vehicle, after several drinks.

Kelly had three different systems. California was not a market for key systems due to Pac Bell's low monthly rent for six-button sets and twelve-button "Call Directors" at $2.00 and $4.00 each. However, the market for small systems in other states where the same telephones rented for $17.50 to $20.00 a month was very good. One of Kelly's telephones resembled the head of a light green alligator. Another, from Nitsuko, offered potential. We placed an open order with Kelly allowing him to sell to the Arcata offices directly with National Account favored pricing. Shortly thereafter, Kelly changed the company name to Telephone Interconnect Equipment (TIE) and began setting up distributors and importing product. More about Kelly and TIE continues later.

YOU BET YOUR JOB: RISKS OF BUYING A PRIVATE PHONE SYSTEM

We had learned through our consulting efforts at CCI that getting a customer to break away from the phone company was like separating a child from his mother—the apron strings to Ma Bell were drawn tight. In the sales process, we had to let the phone company make any presentation or recommendation first—even if they didn't recommend any change. And we needed to know the advantages and limitations of the Bell's Western Electric equipment better than the Telephone Company marketing representatives. Even so, we often found some customers calling the Bell rep back to ask their opinion of our proposals. We made our presentations to the executives in the company who could make a decision between competing products. When the decision was delegated to a subordinate to approve our contract, the subordinate was often threatened by his or her boss with loss of his job if our proposed system failed to work out. The first year in business, my company experienced several customers who tried to back out of their contract for a new telephone system— initiated primarily through subsequent negative input from the phone company rep upon learning of the customer's decision.

Only a few private telephone sales had occurred in the U.S. in early 1969. These were justified more on price than features and consisted primarily of Japanese systems. Still, these early systems had features that Western Electric lacked: the ability to transfer an outgoing call, flash hook hold, and three-party-conference. Music on hold was also an easily added feature. Western Electric telephone systems could have included these features, but there was little motivation in developing them, as they added no revenue for AT&T. If an outgoing call could not be transferred to another internal party, then another call had to be generated, increasing AT&T revenue. Of course, multi-button telephones could be added by the subscriber at an additional cost which would provide access to the same PBX extensions, allowing any incoming or outgoing call to be transferred— again, increasing the equipment rental revenue for the local Bell Operating Company. With all of Western Electric key telephones, any other person in a company that had access to the same business or PBX line could hop on at any time to listen in. Great for conferences, bad for privacy, as many embarrassed executives found out.

THE SALES PROCESS

The sales team in more successful Arcata branches focused on a consulting approach rather than selling hardware to reduce telephone expenses. First, we identified and secured an appointment with a qualified business subscriber—ideally, one that had outgrown their system or was in the process of moving. After obtaining a "Letter of Agency" that enabled us to work with the phone company, we would obtain the subscriber/client's current telephone inventory record and conduct an audit. During this phase we would also analyze the way customer calls were handled—identifying any bottlenecks, cost issues, growth patterns, customer support, and problems. The local telephone company would be invited to submit a telephone proposal for the client that addressed the client's objectives and any issues. After analysis of the Bell proposal, we would make our presentation. We reviewed the client's objectives and problems as well as how the Bell System recommendation did or did not address those areas. The Bell System installation costs and reoccurring monthly equipment rentals were estimated over a ten-year period. Our presentation included our recommended "private" telephone system and solutions to the client's customer calling patterns and problems. After the subscriber/client agreed that our recommendation was superior to that of Bell's, we then reviewed costs. Compared to the Bell System charges, in most all cases, a private telephone system could be justified by an outright purchase or lease. Most customers chose to lease or rent the system as they had been used to paying monthly rent for the Bell System equipment, and the operating cost of a private system was less.

The San Francisco district office of Arcata seemed to be creating most of the initial momentum. We had more than tripled the sales force and outgrown our little carriage office and next-door annex on Filbert Street. I found an old two-story office and spice warehouse on 755 Davis Street in San Francisco and had it sandblasted and renovated. Obviously, we needed a new telephone system and chose an Ericsson 741 PBX. Instead of housing the PBX switching cabinet in a closed room as was the normal practice with Bell, we chose to display it in the stairwell—encased in glass. Colored lights mounted on the side were connected to relays on the racks to indicate calls going through the system. It was an impressive display in 1970. A subsequent Arcata customer, Fred S. James, Insurance, liked the idea so much they had a similar display installed in the reception area of their San Francisco offices.

With initial sales came some initial problems. PCAs, or protective connecting arrangements, were not always available. This caused delays in customer moving plans and placed an added burden on those selling and installing private telephone systems. Some customers were told by Bell System reps that if they had ordered a system provided by the Bell Operating Company, there would not have been a delay.

I decided to take a new salesman, Frank DeFoe, to a presentation for a firm that was growing rapidly and needed a new telephone system for their move to larger space. I'd already conducted a review of the client's needs and a survey. DeFoe and I spent about an hour reviewing that survey, our presentation process, and each page of the presentation binder. I emphasized the importance of receiving an acknowledgment or agreement from the client before moving to the next page. The client had been prepared that our presentation would take about an hour and half. The firm was Rolling Stone magazine. The president was young, dynamic Jann Wenner. After introducing DeFoe, we were invited to begin our presentation. Placing the presentation binder on Wenner's desk, I started to review the last meeting with Wenner and the conclusions to which we agreed. Wenner grabbed the binder from my hand, flipped through the pages, and pointed to the page of our recommended system. "I'll take that one," he said. He signed the contract and directed his financial VP to provide us a deposit check. It took less than three minutes.

On the way out, DeFoe said, "I can do that. Selling telephone systems is quite easy!" I had to explain that it was the first time I had experienced a customer acting that quickly.

There is a little more to the Rolling Stone story. It was our first Ericsson 561—much larger capacity PBX than the Ericsson 741, and much larger in physical size. It was delivered just in time for Rolling Stones' move in. However, the PBX would not fit through the front door—or any door. We had to remove a second story window and rent a crane to get it inside. We were very lucky that the Bell PCAs were installed, and we cut the system in over the weekend. Training on the system commenced the following Monday.

The Los Angeles Arcata office had several prospective customers but had not yet been able to land a contract. They asked to fly two of these customers up to see the system in operation. Rolling Stone graciously extended an invitation, and the date was set for the following week. Three executives from the two prospective firms arrived on time with three representatives from the L.A. Arcata office. The group optimistically opened the doors to the reception area. Behind a very pleasant receptionist was a large eight-foot-long banner hanging on the wall. Written in two-foot-high letters was the welcoming greeting: "THE FUCKING PHONE SYSTEM DOESN'T WORK!" While the L.A. Arcata personnel were trying to save the situation, we went into action to determine the source of the problem.

Almost all private PBXs had some additional features with which Bell business subscribers were not yet acquainted. One was a "switch hook flash" feature that enabled access to other features. The "Flash" feature is available on almost all home and business telephones today—usually accessed by a dedicated button. With early private systems, the standard telephone did not have a dedicated button for the flash feature. The user simply depressed a hook switch (buttons in the receiver cradle) for a moment placing the current caller on temporary hold. This allowed the subscriber to do any of the following: receive another incoming call, transfer the call, make an internal call, dial an outside number, or, if they depressed the flash hook a second time, create a conference between two or three parties. If no action was taken and the receiver was returned to the cradle, the system would wait a few moments before ringing the subscriber's telephone to remind him the caller was still "on hold."

Obviously, in Mr. Wenner's case, we needed to do some additional training. Earlier in the day, he'd had an extremely angry conversation with an outside party. Reaching his limit of frustration, Wenner told the caller he would never speak to him again and slammed the receiver down on the telephone cradle. The receiver bounced enough to create a "flash" to the system, placing the "never-to-be-spoken-to-again" caller on temporary hold. After a few moments, the telephone rang back, and Wenner, picking up the receiver, cheerily greeted the irate caller who had been on hold with a friendly "Hello." Then, he repeated the process with the same results. It was too late to explain the problem to the prospective buyers from L.A.—they were already on the way back to the airport.

Frank DeFoe became one of our top salesmen in San Francisco. I decided to open an office in Denver later in the year and DeFoe was tasked to manage it. While looking at office locations in the Denver Tech Center, we had driven to a third location when the realtor announced, "The other Arcata VP really liked this one." It turned out that as I was expanding into Denver, my counterpart, the Southwest Regional Arcata VP, Jim Villas, had been there a few days earlier meeting with the same realtor attempting to open an office. Some of the acquired VPs were used to making decisions without consulting anyone.

After installing the first few private telephone systems and receiving good recommendations from the customers, all Arcata Communications offices began to enjoy increased sales, and the company started to see rapid growth. Observing the growing success, Arcata National VP, Don Thompson, exclaimed in an executive meeting, "Arcata Communications will be written up in the Harvard Business Review as an example of a successful emerging company in a new industry." Perhaps, the rapid growth was a little too rapid. Expenses began to rise as Thompson seemed to care little of the cost of chartering flights and entertaining prospective sellers to establish new offices, all of which was coming out of Dorsey's budget. Thompson continued to emphasize the importance of expanding offices regardless of Dorsey's protests about a lack of qualifications of prospective sellers. All the Arcata-acquired companies' officers' compensation plans were tied to sales. The greater the sales volume, the more the former owner could earn. All branch offices were placing pressure on Arcata's financial department to approve leases on pending sales. Incomes of the Arcata Communication's regional VPs began to get out of hand—at least in the opinion of Arcata National's Board. Dorsey had proposed Arcata National also purchase Voycall, a small intercom company located in Oakland. Voycall had developed and was manufacturing an attractive "dual phone" that married a multi-button telephone with a hands-free intercom system. Both could be used simultaneously. Neither Thompson nor Arcata's board supported the acquisition. Thompson's and Dorsey's objectives began to conflict.

Don Thompson called asking if I would mind moving from the San Francisco district office to Arcata Communications headquarters on New Montgomery Street. At the time, I was acting as the San Francisco branch manager as well as running the region. I was very happy with the way the building at 755 Davis worked out with views of the Embarcadero and Bay Bridge. I also enjoyed being close to the action and participating in sales meetings. My respectful decline of the offer wasn't enough. Apparently, the company had a problem with a VP—John Barbour in Chicago—and as he had an employment contract resulting from Arcata's purchase of his NCPS, they needed to place him somewhere. Reluctantly, I agreed to relinquish the San Francisco office and move to a smaller office at headquarters. I met with Barbour and told him there were two conditions he was to honor: First, he was not to hire Dave Martin, the unscrupulous salesman. Second, he was not to hit on my former secretary, an attractive, religious redhead with green eyes and a great figure. She was very professional and excellent at her work. When hired, she explained the reason for leaving her last job was because her boss had continued to make sexual advances and comments to her. Barbour was enthusiastic about being back in the Bay Area and readily agreed. The move over to New Montgomery Street was complete. I had finished putting things away in my new office when the phone rang. It was my former secretary. She informed me that she'd just quit. Her attractiveness was much stronger than Barbour's moral compass.

Friday afternoon, I went back to my old office, thinking I might join the sales team for a drink. Victoria's Station was across the street, and it was not unusual for a few of us to head over after work for a cocktail. Walking up the stairwell at 755 Davis St, I was admiring the installation of the Ericsson 741 PBX cabinet and its flashing lights. Sounds of a sales meeting in full swing reached my ears at the top of the stairs. Good—that meant that all my old sales team would be there. I was in a very good mood until I saw the person conducting the sales meeting was none other than Dave Martin. Barbour's moral compass wasn't just weak—it was completely missing. No one had told me what caused Barbour's re-location to San Francisco, but I didn't need to be told.

While Barbour was an extremely creative salesman, he wasn't without detractors. Pat Howard had worked for Xerox for five years before he decided to make a change and sought a position in the Interconnect industry. He would excel, later becoming CEO of Compath, a major California Interconnect company and ROLM distributor. His first introduction to this new industry was being interviewed by Barbour. Howard relates, "As compared to my experience with Xerox and those I knew at IBM, I thought of him as most unprofessional for a VP. And thought to myself, 'if this guy can make it as a senior manager, I certainly can.'"

Don Thompson held a meeting at Arcata National's head-quarters on Sand Hill Road in Menlo Park. All VPs were present. Dorsey, never late, had not arrived. Thompson began the meeting. I interrupted, asking if he shouldn't wait for Dorsey. He shot a quick glare in my direction and continued, stating that Dorsey had been terminated. Shortly thereafter, I had been tasked with doing a competitive analysis of the telephone equipment tariffs in three potential new markets. Upon my return, Thompson held another meeting with all the Regional VPs. He announced there would be a re-org and all regional VPs positions would be changed or eliminated. Some VPs might be offered lesser positions at reduced salary plans. Many would be let go and paid out according to their employment contracts. The business would continue to be run by Thompson and Charlie Smith, the new President for Arcata Communications.

After the meeting, all the Regional VPs and the National Accounts VP made a hasty retreat to the Velvet Turtle, a restaurant on the peninsula, to discuss our options. Around the table sat all Chiefs—no Indians. Everyone talked about resigning and had an idea on how to build a new national Interconnect company. Some claimed access to private financing, others had contacts with venture capitalists. All thought a company could be formed overnight, and all felt they had the qualifications to be CEO of the new venture. Barbour was one of the most vocal about his ability to gather financing for the group. He'd managed to hang on to his position running the San Francisco district for a while. I held no such illusions and had been offered a position by Arcata to build a new "super region" in the Cleveland/Youngstown area—at a greatly reduced salary. Concluding very little would happen in the next month, my new wife and I decided to take a vacation in Europe for the next several weeks and would advise Arcata of my decision upon my return. The thought of moving from San Francisco to Cleveland wasn't appealing—especially with a reduction in salary. Maybe the group of self-assured former Arcata VPs would create a new company in the interim. Within the next year, Arcata was sold to Stromberg-Carlson, a subsidiary of General Dynamics.

THE BIG STEP—STARTING TSI

Returning from Europe, I immediately called my former peers at Arcata anxious to find how the newly formed "National Telecommunication Company" was progressing. It wasn't. Nothing had happened during the month I was gone. Only one regional VP had made a decision. Darrel Nelsen, who had responsibility for Boston and was recently promoted to New York City, elected to start his own company, Natcom, in New York. Nelsen's decision gave me the incentive I needed to strike out on my own. However, the suppliers we had contracted with Arcata were unable to sell product to me in California due to their exclusive agreements still in force. The competitive tariff analysis I had just completed demonstrated that St. Louis and Minneapolis/St. Paul posed excellent opportunities for opening a new office. I shared my plan with my wife and sold the Maserati the following day. With funds from the sale of the car, I flew to St. Louis that weekend to start a new company, telling my wife I would send for her and our dog—a 130-pound Great Pyrenees—in a month. It was the last week of February 1972. Upon waking in St. Louis, Sunday morning on the 26th, I walked from my hotel room to get coffee and a paper. The headlines told of an official from the Steam Fitters Union who was blown up in his car the night before. Some months later, an executive from an independent telephone company had a bomb go off in his car, shattering his legs. It was concluded the bomb was intended for another individual who drove a similar car. What sort of a place was this to start a business? That kind of stuff wasn't happening in San Francisco and Marin County where we lived. The Tenderloin area and the Barbary Coast in the city could become rough at times, but for the most part, an aura from the '60s lingered over the Bay Area—"Free Love" and peace signs prevailed. Now, in St. Louis, "the Jeweled City by the

Bay," as San Francisco columnist Herb Caen referred to it in the '60s, was far away.

I found a small eight-by-eight office adjacent to an answering service and made arrangements to rent the office, subscribe to the answering service, and start making appointments. Both my attorney and accountant in San Francisco set up introductions in St. Louis. I chose the name Telecommunication Services, Inc. (TSI) and filed incorporation papers by the end of the week. Trying to get business started in the early months turned out to be very difficult. The first few sales would not be easy, nor did I think they would be. What I hadn't realized was that businessmen and bankers were more conservative in Missouri. The CPA in San Francisco who provided advice on my business plan had lived in St. Louis. He told me, "People sleep on their money there." Now I understood what he meant.

When informing businesspeople that they now had an opportunity to select telephone equipment from others than the local telephone company, their reaction was one more of distrust and disbelief than interest in obtaining improved features to serve their customers or lowering the firm's telephone expenses. In confirming a local address, I often mispronounced the French names for streets or communities, a definite clue that I wasn't from the area. The only thing positive in those first few months was the arrival of my wife and our dog. The airfare cost more for the dog. My wife, Susie, had a pleasant surprise for me—she was pregnant. We rented an apartment in Creve Coeur, west of the downtown area. One morning while driving to my "answering service office," a new gleaming white van passed me. On its side, in bright red letters and catchy logo, read "ITS—Independent Telephone Systems." Uh Oh! In my competitive analysis, I'd missed that competitor, but I soon learned that ITS was very competent. If they came across a potential customer I was working on, they were quick to discredit me. At the time, TSI had no sales, no technicians, no references. ITS told my prospects that I was some guy from the west coast and would probably be gone in a month. Competing with Southwestern Bell was going to be tough enough, but local competitors were, in the beginning, much tougher.

CHAPTER 7

Microwave Communications, Inc. (MCI): Competing with AT&T Long Lines

The following company did not engage in the sale of private telephone systems. However, its encounters with AT&T parallel those experienced by Interconnect distributors.

Competition to AT&T came in two forms. In 1968, the FCC—responsible for the regulation of the communications industries, including telephone service, and radio and television broadcasting, had ruled that the terminal equipment market should be opened up for the first time to others beyond AT&T. Virtually all terminal equipment or customer premise equipment (CPE) was owned by AT&T and rented to its subscribers. This included every residential telephone, business telephones, and switchboards. When AT&T's service began deteriorating in New York and other major cities, pressure to act was placed on the FCC by the nation's largest corporations, most of whom where headquartered in NYC and extremely unhappy. One year after the FCC decided that private telephone systems, mobile radios, and answering machines should be allowed to connect to AT&T's network, the FCC made a second major decision which rocked AT&T's headquarters. The commission granted the application of a small company trying to offer a private line microwave service between cities. MCI, or

Microwave Communications, Inc., was attempting to offer intercity service between Chicago and St. Louis.

Initially, the intent of MCI's service was to afford companies with offices in both cities the opportunity to connect to each other at a lower cost. Upon dialing from one city location to the distant city's location, a company's calls would be switched at the Bell Company's central office to MCI's microwave towers and beamed to the other city. MCI's founder, Jack Goeken, had filed an application with the FCC five years earlier to provide point-to-point private-line service over microwaves between the two cities. Goeken, short of cash, was introduced to William McGowan. McGowan, a scrappy individual who worked his way through Harvard Business School, was investing in new ventures with cash resulting from the sale of a company. At the time, MCI had no full-time employees or finances and was even smaller than Tom Carter's Carter Electronics. McGowan decided to invest, becoming the company's chairman and CEO. Fortunately for him, a year later in 1969, the FCC approved MCI's application.

In order to presell its service to customers, MCI needed to negotiate "local loop" connection rates. The loops are the Bell Operating Company lines between the customer and the AT&T central office switch affording connection to the distant city. AT&T had fought hard to prevent the FCC from granting approval, but once it did, the FCC did not provide any guidelines for establishing the rates they could charge for such connection nor did they provide any timeline for performance. MCI initially thought the monthly rental for the lines from the customer to Bell's central switching office would be comparable to the rents normally charged to their business customers for dedicated circuits. AT&T thrived on lengthy negotiations—except for pending rate increases. The longer AT&T could stall, the less the impact of the final settlement would be to the company. Meanwhile, the competitor risked running short of cash or customers while AT&T dragged its feet.

Negotiations between the two were going badly. AT&T had changed its mind and had decided instead on a "capital contribution" plan in which MCI would be required to help pay for the maintenance and upkeep of AT&T's nationwide network—the miles and miles of expensive wires and cables.

In February 1972, John deButts became the new chairman of AT&T. He had invited the presidents of Bell Operating Companies to a week-long conference in Key Largo, Florida, to hear their views and complaints. DeButts, after graduating from Virginia Military Institute, had worked his way up the ladder in virtually every area of AT&T. He was determined to improve the quality of telephone service that had been declining in major cities. As a corporate figure in his fifties, deButts was imposing; standing at six feet two inches and 200 pounds, he exuded leadership, power, and strength. He also spoke in a slight southern drawl.

DeButts listened to the complaints of the operating presidents who were focused on losing both terminal equipment and their long-lines business. MCI had gone after some of their largest customers. The presidents felt AT&T's New York management had done little to protect their markets, and they wanted to stop competition now. T.S. Nurnberger, president of Northwestern Bell exclaimed, "I would meet 'em or beat 'em. You bastards [the competition] are not going to take away my business!"

Most presidents at the meeting felt AT&T needed to act decisively and quickly before MCI and other competitors had established businesses that regulators would prevent AT&T from dislodging. Nurnberger added: "How many MCIs will proceed with construction plans if we file matching rates now? A big fat zero!"

Charles Brown, president of Illinois Bell and the first target of

MCI, chimed in: "There are large amounts of revenues that are vulnerable, which we can preserve if we choke [them] off now. I think you have to hit the nails [competition] on the head."

Later, when Brown replaced deButts, he would come to regret the comment. One year after the Key Largo meeting, John deButts announced in a press conference that AT&T would depart from its long-term strategy of nationwide average pricing so that it could more effectively respond to new competition. After MCI's successful $100 million stock offering, deButts decided that AT&T wasn't going to let MCI get away with "cream-skimming." MCI had been conveying to its prospective customers a $100 per circuit per month savings over AT&T's private line services. AT&T would begin implementing a new pricing for its long-distance services called "Hi/Lo." AT&T promised that a large portion of their customers would see their rates reduced by thirty percent. The reduced rates for AT&T private line service on the same routes proposed by MCI would effectively wipe out MCI's proposed savings to its customers.

Seething mad, McGowan met with deButts in AT&T's corporate offices. The meeting did not go well. McGowan threatened litigation, but deButts was not about to be intimidated. After the meeting, McGowan decided to embark on a three-pronged strategy. He would have to convince the FCC, Congress, and the courts that AT&T's anticompetitive actions warranted a complete restructure of the company or a federal antitrust suit. However, John deButts was developing a strategy of his own to choke off competitors.

CHAPTER 8

AT&T's Additional Barriers and Dirty Tricks

AT&T was waging both legal and regulatory battles against competition in the long distance market while the local Bell Operating Companies waged their own kind of war by passing regulatory and legal processes.

O n the West Coast at Arcata, we encountered only a few mishaps coordinating installations with Pacific Northwestern Bell, Mountain States Telephone and Telegraph, and Pacific Telephone and Telegraph (PacBell). The installation of private telephone systems was new to both the providers and the Bell System Operating Companies in the late 1960s and 1970. Nothing at the time indicated a pattern or deliberate noncompliance. However, coordination between Southwestern Bell (SWBT) in Missouri and our small company was going to be a new experience. Chuck Schmichel, from Arcata, arrived to become TSI's operation manager. He had been offered an equity position in our new company. We'd wired our first four telephone systems sales and installed, tested, and prepared the central exchanges (KSUs) for cutover. The customer agency letters had all been filed with Southwestern Bell Telephone, and advance notice had been given to SWBT of when the connecting arrangements (PCAs) were needed for each.

SWBT failed to acknowledge receipt of any of the agency letters.

At the first customer location, Bell technicians did not arrive when we'd asked them to. None of the PCAs had been installed, and SWBT hadn't explained the delay or asked us to reschedule. The same thing happened at the second location. We followed up with letters and left phone messages to Bell contacts. No response. The next two customers were scheduled for the following Wednesday. The first two installations were moved out to the end of the week. Despite repeated communication, at the end of the week, none of the SWBT connecting arrangements had been delivered or installed, and we had not been provided with any communication by SWBT indicating reasons for the delay or when to expect the PCAs. Our customers began to have second thoughts about the decisions they'd made for a private telephone system. Our small company's credibility was on the line. It was embarrassing. Three of the new customers were understanding; the fourth wanted his deposit back. Knowing our customers and our small company could not continue to tolerate SWBT's tactics, I anguished over what to do. Over the weekend, I provided instructions to our technicians and sent a telegram to the legal department of SWBT in St. Louis:

> To Southwestern Bell Telephone Company, Legal Dept.
>
> Our company, TSI, has contracts with 4 St. Louis customers who have been waiting for the completion of the installation of their private telephone systems. Regardless of repeated requests for connecting devices (PCAs) and written advance notice of scheduled cut-over dates, SWBT has failed to provide the devices and has caused unnecessary costs

to our company and inconvenienced our
customers, denying them the benefits of
their new systems. To complete our
contracts, all four systems have been
"hard-wired" to your network trunks. We
will provide SWBT notice of our intent to
continue this practice with copies to the
FCC until such time as a reasonable
response to our requests is provided
by SWBT.

Paul D. Olsen, President

Early Monday morning, I received a call from SWBT's legal department requesting an immediate meeting. The connecting devices had all magically appeared, and we rewired the customer trunk connections to the PCAs to make the installations comply with the FCC tariffs. SWBT stated they would cooperate with our requests in any new installation. At least, that seemed to be their intent. Delivery of the PCAs improved, but anticompetitive activities began to show up elsewhere. St. Louis wasn't the only area where problems of PCA deliveries and their failures existed. A pattern by the Bell System began to grow.

About the same time, Jerry Burns, Arcata Communications Branch Manager in Orange County, had three or four installations needing PCAs. A few had been installed at one customer that was not moving, and PacBell's installation of PCAs was not time critical. However, a major customer of Burns, Hadley Hardware, had contracted for a private system and was moving to new offices. Burns tried several times to get PacBell to deliver the necessary PCAs but received absolutely no response from PacBell. To complete the contract on time, Burns requested his technicians remove the PacBell PCAs from another Arcata customer location that was scheduled for a cut-over later and install them at the new Hadley Hardware offices in time for their cut-over. A few days later, PacBell found that Hadley was making calls from their new location and visited the site to discover the system was connected to PCAs from another location. PacBell threatened Burns with arrest and told him: "We'll have you thrown in jail for stealing PacBell property if they aren't returned immediately."

Burns reluctantly complied, connecting each of the Hadley customer trunks to single line telephones until the PCAs were installed a week later. Fortunately, the customer understood the fault was not Arcata's but PacBell's and became a good reference for Burns.

Litton Business Telephone System, a later entry in the Interconnect market, eventually took legal action against AT&T. Contained in Litton's testimony in their antitrust suit against AT&T, Bell Operating Companies refused to acknowledge the receipt of Letters of Agency arranging for cut-over dates, arbitrarily changing cut-over dates, and failing to provide PCAs. Litton further argued that PCAs occasionally malfunctioned, causing actual injury and technical insult. Problems were compounded when a PCA malfunctioned.

Roger Williams had been in financial management with General Electric before joining Arcata Communications. After the acquisition of Arcata by General Dynamics, the company concluded an office in San Diego was not needed and asked Williams to close the branch. Williams took advantage of the situation and opened his own Interconnect company. I asked Williams if he encountered problems with delivery and operation of PCAs.

"Oh God, Yes!" he exclaimed. "PCAs had a high degree of failure. When a failure happened in the protective device, the trunks would no longer hunt."

What does this mean? Here's an example. A business has a bank of ten trunks that are programmed to hunt. Incoming calls are routed to the first trunk (commonly the main number for the business). If that trunk is busy or in use, the call "hunts" to the second trunk and continues until a vacant trunk is located. If all are busy, the caller receives back a busy signal. Outgoing calls usually hunt from the last trunk. If the PCA on trunk number three has failed, the hunting for the incoming call stops and does not hunt to an available trunk—restricting incoming calls to no more than three.

Williams lamented further: "The customer calls in a service call, and we go out and trouble shoot the system. The system tests just fine, the problem appears to be Bell's PCA. We wire around the device, connecting directly to the Bell trunk—and the system works fine. We just spent money responding to a service call that was not our problem."

In Houston, Jerry Burns' twin brother, Jack, had his own Interconnect operations, an extension of his brother's company. He, also had similar issues. "We had a lot of problems with connecting arrangements," Jack shared. "I'd finally tell the lead installers when they encountered a problem with PCAs: Just by-pass it." This meant connecting directly to the SWBT trunk lines. Jack Burns claimed that he never got a call from SWBT complaining about his installers' wiring directly into the network.

AT&T's connecting arrangements were not the only problem. When they were installed and *did* work, they still added a cost burden to the Interconnect companies trying to compete with AT&T as well as adding unnecessary costs to the customer's monthly telephone expenses. Regardless, we still managed to win sales. Our small company, TSI, began to grow. We managed to hire some of the most effective salesmen who had been competing against us, moved into new offices, and leased a fleet of service and installation trucks.

Soon we began to find other installation obstacles placed in our path.

BELL OPERATING COMPANIES' DIRTY TRICKS

Selling a phone system to a customer at an existing location provided an opportunity for installers to use the existing cabling or wiring. Bell claimed "inside wiring" was their property, but we were allowed to make an offer to purchase it—assuming the wiring was compatible with the telephone system chosen by our customer. Regardless of the age of the old Bell wiring, though, they quoted us a price that exceeded what our cost would be to install new cabling given equal access to the customer's conduits. Bell's anticompetitive wiring practice began to surface at all existing locations where the customer was not moving. We took another tack. We would wire around Bell's old cabling. This was more expensive for us, though, and didn't always result in an attractive solution as a lot of the low voltage wiring would be exposed.

When we declined to accept Bell's outrageous price for their old wiring, long written off, we requested Bell remove their cable so that we could use the customer's conduits. If they were unable to remove the cable without destroying the customer's premises, we requested Bell abandon the cable for the customer and our use. Sounds logical, right? Bell came up with another anticompetitive tactic. Soon our installers began to report back that they could not complete the installation because Bell employees had chopped the cable off behind the multi-pin connector to which we could have connected our telephones, rendering the cable useless. It would take hours for our technicians to rewire a twenty-five-pair or fifty-pair pin connector to the old wiring—assuming we had enough cable length to work with.

Bell employees went further. We began to see customer cabling chopped off at the wall or conduit. Frequently, a broom handle had been used to push the severed cable several inches down into the conduit so neither the cable nor the conduit could be used. While completing one installation, our cartons of cables, some telephones, and installation tools left at the customer site were taken. We reported all the above instances to SWBT and our national association, The North American Telephone Association (NATA), that several of us had created in the early '70s. NATA, located in Washington, D.C., was our industry contact point for AT&T and the FCC. We began to document instances where we felt AT&T was employing unfair and unethical competition tactics.

Other minor nuisances came in the form of switching incoming calls to other locations. TSI sold a large TIE key system to a baby shoe manufacturing company in St. Louis. On the day of "cut-over," incoming calls to the shoe manufacturer were routed to a beauty salon, and calls to the salon were routed to the shoe manufacturer. This happened in two separate instances, then never surfaced again. As switching the lines between customers could easily be done at SWBT's central office, some Bell employee probably thought it humorous. However, neither customer did. The incidents only created doubt in the customers' minds as to whether they made the correct decision in breaking away from Bell. The president of the shoe company asked for my personal assurance that it would not happen again.

We began to see additional tactics employed by SWBT marketing people. When confronted with a possible loss of a business subscriber, Bell employees would tell our prospective customers, "Don't make any decision now, Bell will soon be introducing a brand new product at a lower monthly rate." That preemptive marketing tactic had the effect of delaying, if not completely stopping, our sale. The tactic was conducted before the so-called lower monthly rate program had been submitted to and approved by State Public Utility Commissions. Again, any tactic AT&T employed that would delay a sale benefited them by continuing their current equipment rentals. The following year, several, if not all, Bell Operating companies began to market "Two-Tier" pricing for some of their business telephone systems. This was AT&T's attempt to compete against Interconnect companies that offered a fixed lease rate on private systems. The "Two-Tier" program included a portion (roughly half) of the monthly equipment rental at a fixed rate while the balance would be subject to periodic increases as approved by the State PUCs. Bell did not sell or lease their telephone systems, but the two-tier program provided them a competitive tool. The program still didn't offer much in user features, but the tiered tariffs complicated the sales situation.

• • •

BOOK TWO

National and Regional Interconnect Companies Emerge

Along with Arcata Communications, other firms in the early '70s recognized the opportunity to gain a nationwide presence in the new emerging market. Some succeeded, many did not.

CHAPTER 9

Universal Communication Systems (UCS)

The Krisch family—brothers Joel and Adolph along with their sister Rosalie—had established a chain of more than sixty-five Holiday Inn Hotels and took their company public under the name of American Motor Inns. Early to recognize the benefits private telephone systems could bring to their properties as well as all hotels and motels, they formed UCS. Fred Shaftman, son of Rosalie, became its President. UCS hired Bob Goldstein who had been working as a garment salesman in NYC. As UCS's CEO, Goldstein knew how to contain costs and maximize profits. Former employees thought of him as tough and sometimes mean. As some described it, "Bob could 'go off,' verbally abusing an employee—to the point you thought you were about to be fired. Then he'd say, 'let's go have dinner and some drinks,' like he was your best friend."

There was no question Goldstein ran a tight ship to keep costs down. Wallace Hammond, UCS operations VP, tells of how one day Goldstein went through the Roanoke, Virginia, headquarters and seized all the pens off each desk, leaving only one. He then circulated a memo stating that anyone needing a pen should come see him. According to former UCS Regional VP Peter Campbell, "He not only did that with pens, but with paper clips as well." Campbell further added that when he was sent out to California to help "clean up" the West Coast operation, he found that the technicians had no air conditioning in the Riverside office where it gets very hot in the summer. The company didn't want to pay for it. Taking the initiative, Campbell had the AC installed which improved morale with the technicians and installers. Even Goldstein admitted, the installation guys were never happier.

As an example of the company's frugality, during the installation of a new phone system for a customer, the UCS coordinator came to Campbell's office with phone parts that were rusty, and plastic covers that were ill fitting and scratched. Instead of new telephones, headquarters had shipped out low-cost refurbished ones which had been purchased from a reseller of used telephone equipment. As Jason Horrell, former UCS sales engineer and customer problem-solver, stated, "It was a love/hate relationship with Goldstein. He had lots of charisma and was very smart—he loved to entertain people with his yo-yo tricks. Goldstein personally approved all flights and bought the tickets for regional installation managers and salesmen. Installation personnel did not fly." All sales proposals including the system pricing were prepared at Roanoke headquarters and mailed to the salesperson to deliver to the customer. Hammond added, "Management seemed to trust no one and ran the business [with] their thumb on the neck of everyone." Goldstein was quoted as saying, "It's a sad-sack business to be in. The only way we can operate [is] if we [management] have our hand on everything."

After the sale was made, the customer site would be reviewed by one of the Roanoke sales engineers before beginning the installation. UCS developed a strong technical ability. They would buy components from NEC and Hitachi, assemble crossbar racks, configure the PBXs to customer specifications, and complete testing in their Roanoke facility before shipping to the customer site—which could be anywhere in the U.S. UCS established regional offices in most major cities. Their facilities were spartan with little overhead. More traditional offices in the Interconnect industry might include a branch manager, sales manager, salespeople, office personnel, accounting, warehouse manager, and local service and installation technicians. UCS developed a different model. Installation personnel and service people would be dispatched to where they were needed from Roanoke. Many worked from home. Salespeople were encouraged to rent space in shared executive suites. While other Interconnect companies focused on professional firms, auto dealerships, banks, and small companies, UCS's primary markets were hotels, colleges, and VA Hospitals. With their low overhead and ability to configure their own PBX systems, they were often the lowest bidder. A monthly newsletter, "Inter-Connection," in November 1974, reported key features of UCS managerial foresight and operating flexibility as:

- Equipment used was almost exclusively Hitachi and assembled by UCS (NEC components were later used).

- Initial operations were concentrated entirely in hotel and motel systems before expanding into business installations.

- Geographical area to be covered was carefully selected. Initially this was the East Coast, then the West and South.

- Maintenance control was centralized. Maintenance records were maintained on UCS's 1000-plus installed systems at headquarters. In-bound WATS numbers were used by customers to request service.

- A separate quality control department of five men plus an installation supervisor kept trouble calls to a minimum

- Attention was given to cost savings (an understatement) as 126 installation maintenance people worked out of their homes. In some cases, installer's wives traveled with their husbands to train operators on how to use the new systems.

The company was eventually purchased by Prime Motor Inns for what several UCS managers and the former Senior VP of Finance, Dallas Jarrell, believe was about $80 million. Jarrell commented that he thought Bob Goldstein was the most successful CEO in the industry at the time. Goldstein died the same year PMI acquired UCS. The above individuals also thought that PMI later sold UCS to Bell South after the break-up of AT&T. Estimated revenues at the time were $120 million to $150 million. Unfortunately, when I inquired through the Freedom of Information Act and the Securities and Exchange Commission (SEC), no other information was found. A similar request was made to the National Archives and Records Administration. NARA was unable to respond, as its operations were closed due to the COVID-19 virus.

CHAPTER 10

Telecommunication Systems of America (TSA): Part I

After college and a stint in the Air Force, but before joining Teleci as their first salesman, David Perdue had been selling corrugated boxes. He had read about FCC's *Carterfone* decision in the *Wall Street Journal* and put it aside thinking it would probably be a good opportunity for someone. Not too long after reading the article, he was contacted by a head-hunter who had set up an interview for him with Teleci. Teleci was a small company that installed telephone systems and central offices for independent telephone companies. Some 1,500 independent telephone companies usually followed the same practice of subcontracting out their telephone system installations. With the *Carterfone* decision, Teleci also saw an opportunity in the sale of private phone systems. Perdue was tall, courteous, assertive, and professional. Teleci decided that his qualities, along with his strong sales ability, outweighed his lack of industry knowledge. They made him an offer that included a salary plus commissions, worth much more than what he was earning at the box company. With a wife and three young kids, Perdue accepted.

To provide him with the knowledge he would need, Teleci hired a retired sales manager from South Central Bell. Working eight hours a day for ten days isolated in a hotel room, the retired Bell manager taught Perdue everything he needed to know about the Bell System, Western Electric's phone equipment, and its tariffs. Teleci soon found U.S. telephone manufacturers would only sell to Bell or Independent Operating Telephone Companies. In the South Central Bell area, like California at the time, the smaller private key systems were not competitive to the Bell offerings. The additional cost of protective couplers made it a nonmarket for Teleci. They would need to focus on the larger PBX market.

Perdue lamented to me: "Initial sales were a real struggle. In the first few years, only Japanese systems like OKI, Nippon (NEC), and Hitachi were available, and all the brochures and technical materials were printed in Japanese. When a prospective customer asked if he could see a system, all I could do was point to a picture on the brochure and tell him, 'Here's what it looks like.' Then they would ask, 'Well, who else has one installed?' I would answer, 'None are installed as yet. You'll be the first one!' I think we only sold three or four systems the first year."

Regardless of the difficult beginning, Perdue was accomplishing what Teleci had hoped for, and in 1971, they made him Vice President of Sales. Perdue sold a system to Kemmons Wilson, founder and CEO of Holiday Inns. Using that sale as a springboard, he proceeded to sell several PBXs in various cities owned by Holiday Inn's franchisee, United Inns. Wilson concluded that all Holiday Inn franchisees needed the features and cost savings a private PBX could provide, and Holiday Inn bought Teleci. Perdue added, "Shortly after the sale of Teleci, a VP from Holiday Inns came in and said to me, 'We want all your records and files.'" Perdue protested but acquiesced, surrendering the documents to him. That was all the motivation he needed to decide he would start his own company.

He teamed with a co-worker, John Mitchell, from his box-selling days, and with a few investors raised $40,000 to form Telecommunication Systems of America (TSA). "I developed a short 'pro forma' showing projected sales and profitability to all the local banks in anticipation of securing a loan. I was in my early thirties and naive to the requirements banks needed to secure a loan." The loan didn't happen. Undaunted, in the first six months of TSA's operation, Perdue managed to sell the first private "Centrex" system in the U.S. to the National Bank of Commerce (NBC) for their new headquarters. Selling a large private Centrex system was unheard of in the early '70s. The task would be especially difficult without the credibility of name recognition or a large customer base. Most Interconnect companies at the time would have considered the sales effort futile. "We might have only sold a couple of PBXs to motels before this sale," Perdue stated. He asked for and received a down payment of $250,000—half of the sale's total. Since the sale was made to NBC, Perdue thought it only fitting to open another TSA account there and deposited the funds. Within the week, he received a phone call from the NBC loan officer who had turned down Perdue's earlier request for a loan.

"Dave, is there another TSA company in Memphis?" the bank officer asked.

Perdue replied, "Better not be. We're the only one that I know of."

"Well, someone with the same name just made very a large deposit in our bank with a check from our bank."

"That's me," Perdue proudly replied.

Amazed and somewhat flabbergasted, the bank officer referring to the recent sale to NBC added, "Uh, well ... Congratulations."

A Centrex system is a section of a Bell System central office. Central offices handle thousands of calls with trunks and lines connecting subscriber PBXs, Key Systems, and land lines. The system does not reside at the subscriber's office, but instead sits on Bell's premises. Depending on what the telephone company determines the traffic load requirement is, cables are connected from the central office to the subscriber's location. As such, the customer leases a portion of the central office. The advantages to the customer are that it saves space, provides direct inward dialing to departments and individuals, is easily expanded, and carries more traffic. The disadvantage—it ain't cheap.

TSA also encountered some of the Bell System's dirty tricks. Shortly after receiving the necessary authorization letter from TSA requesting connecting arrangements for the bank's new ITT Centrex-type system, both Perdue and the bank received calls requesting a meeting at South Central Bell Headquarters. Several Bell System executives and attorneys were sitting around a table. They opened with, "We're not going to allow you to interconnect your switch for National Bank of Commerce!" When Perdue requested Bell state their reason for making this decision, they replied that the system, made by ITT whose products met AT&T specifications and had been installed with independent telephone companies, would be injurious to the network—creating damage. This had always been AT&T's excuse for needing protective couplers. Perdue and the NBC bank officer left the meeting. After a few moments of silence, the bank officer turned to Perdue saying, "Look, Dave, NBC doesn't want to get into the middle of a dispute. We'll give you a few days to straighten this out—otherwise, on Monday of next week, we'll have to cancel the agreement and go with a South Central Bell Centrex system."

Perdue immediately chartered a flight to Birmingham and had his lawyer called SCB's legal department. The next step would have been to set up an appointment with the State Public Utility Commission. However, on the call to SCB's legal department, the SCB attorneys said, "There is no reason for you to come to Birmingham for a meeting. Of course we're not going to prevent you from interconnecting your switch for NBC." Bell provided the same information to NBC. To Perdue's relief, the issue was over. But it wasn't the last. A few years passed, and Perdue was able to hire a VP of Marketing from South Central Bell. The man told him that word about a private Centrex being sold had travelled throughout the Bell System. "After it was installed, we got calls from Bell System employees all over the U.S. asking, 'Is it working?'" It was working just fine.

Although Perdue's efforts were needed to help grow the company, he was also a director of NATA, and when requested, would take time off to travel to Washington and lobby for the industry. Perdue; Commander Jim Lovell, the former astronaut and now president of Fisk Telephone Systems; Richard Long, formerly of Fisk and now president of NATA; and Ed Spievak, NATA's attorney, would lobby with various congressmen, including Tim Wirth, an advocate of competition, and Al Gore. The team related AT&T's anticompetitive actions and made the plea for continued support. Perdue made the trip almost every quarter. Why is this noteworthy? There were several of us on NATA's board, me as a Vice President included. When our help was requested, most of us would decline. The travel expense and time away from our companies was our excuse. Perdue offered no excuse.

ROLM (described in Chapter 17) was a new startup company in Silicon Valley that had developed and was successfully selling ruggedized minicomputers to the defense department. In 1973, Dick Moley, the telecommunication marketing director for ROLM, called on Perdue. He visited TSA asking questions about their business and customer installations. Perdue began to work with Moley at ROLM on defining features he felt his customers desired. They met several times before ROLM announced their CBX product in 1974. TSA became one of ROLM's first distributors and made, it seems, the first sale of a ROLM CBX. As sales of the CBX began, TSA discontinued selling ITT PBXs.

During the oil embargo of 1974, TSA sales dropped as much as ninety percent. In the summer of that year, the company needed more cash to keep going. Meanwhile, TSA's bank, NBC, called in the loan, making a demand for repayment. One of TSA's early investors had contacts at the bank and was able to get them to back off while they raised more money. In the process of attracting more capital, Perdue lost controlling interest in TSA. During the same year, however, TSA made another large sale to Phillips Petroleum. The company was based out of Tulsa, but the sale was for several system installations in the North Sea. Almost daily, Phillips would helicopter their employees to the nine oil rigs from Stavanger, Norway—a 200-mile trip each way. Establishing a hotel on one of the platforms saved the company considerable expenses. It also meant that rig's system needed to include hotel features. The oil platforms needed connection to the mainland and between each other. The installation included several switches connected via microwave towers and various alarm notifications. It was a very sophisticated network for TSA.

The following year, Perdue contacted Northern Telecom (NTI), a Canadian company that had spun off from Western Electric years earlier. NTI had established a U.S. office in Nashville intent on penetrating the U.S. market with their SL-1 PBX. NTI first tried selling to independent telephone companies. After their Nashville office had been established for two or three months, Perdue called requesting that a sales contact come to Memphis, meet with TSA, and describe features of the SL-1. A couple of NTI sales reps came—along with their president, John Lobb. Lobb spent some time with the company and seemed impressed with Perdue and TSA, which was doing about $5 million a year in sales at the time. Lobb suggested that Perdue sell TSA to Northern Telecom and invited him to work on a business plan for the U.S. at his home in Fort Lauderdale. He told Perdue that he wanted NTI to grow to a sales volume of $50 million in two years.

David Perdue

Perdue called Ken Oshman, President of ROLM, and informed him that he had a verbal agreement to sell TSA to NTI. Oshman said he would be willing to make an offer, but Perdue had given his word to Lobb and could not entertain another offer. "Oshman was very nice, saying he wished I would have called him first as they would have loved to have done something with TSA," recalled Perdue. Naturally, ROLM cancelled TSA's distributor agreement and appointed another distributor in the area. But that would not be the end of Perdue's relationship with ROLM.

The purchase offer provided an opportunity for the TSA investors to cash out. TSA's Board approved the deal, and the purchase was completed in December 1976. Perdue was made President of NTI's Systems division.

Perdue expanded the NTI sales force, appointed more dealers, and opened direct offices. He exceeded Lobb's sales objectives by reaching $63 million in sales within the first two years and achieved $100 million in the third year. Late in 1978, Lobb retired, and Monty Hayes replaced him. Perdue and Lobb had agreed that the NTI Systems Division would remain headquartered in Memphis, but Hayes decided to move it to Dallas and requested Perdue follow. Reluctantly, Perdue agreed to take his wife to Dallas to look at homes in the area. With three teenagers in school, the couple decided it wouldn't be a good move for the family. Returning to Memphis, Perdue informed Hayes that it was going to be a "No" and subsequently resigned in the fall of 1979. What Perdue did next would be another "first" in the industry.

CHAPTER 11

International Telephone and Telegraph (ITT) and ITT Terryphone Corp.

When the market first opened in 1969, there were only two companies that had a national presence and the technical capability to compete with AT&T for customer premise equipment. Terryphone, owed by ITT, had a nationwide base of offices that had been installing paging and intercom systems. One could claim they had been competing with the local telephone companies. However, their systems more often complemented the existing AT&T telephone systems. Beige telephones looking like standard single line phones were placed, where needed, next to the AT&T telephone on the customer's desk. Each had a speaker built into the face of the phone instead of a rotary dial. Lifting the handset, the user could page through all the Terryphones or selected zones in the business and overhead paging speakers. Another division of ITT also manufactured standard single line telephones, multi-button (key) telephone systems all available in an assortment of colors, and PBXs. The PBXs were primarily made offshore for the international market but were also sold through U.S. distributors. Most all Interconnect distributors bought ITT single line telephones and key telephone systems to work behind any PBX, except where the manufacturer employed their own proprietary telephones that were only compatible with their PBX. ITT single line and key telephone instruments looked just like AT&T's and were similar in function.

As mentioned earlier, in 1925, AT&T was required to divest their interests in European operations. ITT purchased several companies from Western Electric including Bell Telephone Manufacturing Company of Belgium and British International Western Electric. They also bought Kellogg Switchboard & Supply, becoming ITT Kellogg. At one time, ITT was also the largest owner of L.M. Ericsson of Sweden. By the time the Interconnect market was just beginning, in 1970, ITT sales had grown to $8 billion with profits of $550 million. The company had technical experience in pulse code modulation (PCM) upon which future digital voice communication systems would be based.

It is unclear what ITT's long-range business plans were regarding the Interconnect industry. The company was arguably the most qualified to take advantage of the new Interconnect market. ITT had the financial strength and an in-place nationwide marketing, sales, and service organization in ITT Terryphone. ITT sales in the Midwest were not made by Terryphone but by a few scattered sales reps reporting to a separate division. Sales in the Midwest were very modest, but ITT had some significant large customers.

In the mid-1970s, Jerry Pock, responsible for ITT's direct sales and operations in the Midwest, contacted TSI. Pock asked if TSI had any interest in buying their Midwest branches. The deal would consist of little more than assuming their office leases, servicing ITT customers, and hiring their employees—which we did to expand our operations and credibility. The ITT customer base consisted of a few large PBXs and some key systems. No requirements were placed on TSI to sell specific quantities of ITT PBXs. TSI sold a few ITT

Pentaconta crossbar PBX products but found them less reliable than others while offering little cost advantage over emerging digitally based systems. Late in 1977, ITT embarked on developing a digital telephone exchange system, ITT System 1240, later referred to as "System 12." Designed as a central office switch for telephone operating companies, it enjoyed initial strong sales in Europe and Mexico, but the system took longer to integrate. ITT Kellogg in Raleigh, North Carolina, undertook the conversion to the U.S. market, and although sales were announced in 1984 and 1985, with further losses, the project failed in early 1986. ITT divested its telecommunications assets in 1986. Had the project succeeded, System 12 may have been an ideal platform for Interconnect distributors in the U.S. Regardless of ITT's technical and financial prowess, the company failed to achieve a significant position in the Interconnect market place. However, the company or its distributors continued to bid on occasion for large proposals into the 90's.

CHAPTER 12

Executone, Inc.

The other organization with national presence and positioned to capitalize on the new emerging market was Executone, Inc. Based in New York, the company had enjoyed success in most major cities before 1969. Through a network of independent dealers, Executone specialized in nurse-call and location systems employed in hospitals, as well as paging, background music, and mechanics-call systems used in auto dealerships. Executone commanded a major market share in the hospital/healthcare industry. The company's President and CEO was Ed Brody, and his VP of sales was Harvey Hellering. Their business model was to sell products to exclusive dealers in most cities. Dealer agreements contained specific performances required of the dealer including service, inventory, and sales. The authorized dealers, in turn, would sell and install Executone products to local customers in their region. Each dealer had their own sales and installation people. Executone also maintained a direct sales office in New York. The structure appeared ideal to sell and distribute telephone equipment, so the company's management decided that Executone should enter the Interconnect market.

The company didn't realize that this would be a challenge for many of Executone's dealers who'd been in business for several years and were comfortable with the existing line of business. The revenue each received from new sales, moves, additions, and changes with their customer base provided a nice living. The telephone equipment business was a new experience for them, and few dealers, especially the older ones, wanted to undertake the challenge. Dealers would need education and training in the field of telephony—which was virtually nonexistent within Executone's corporate structure. In addition, it was questionable if Executone had the capability to develop or manufacture full-featured, reliable telephone equipment. Their first small key system called "Charter" employed multi-conductor cabling (fat twenty-five- to fifty-pair cables like most early key systems) and was limited in features. Competitive manufacturers quickly evolved from multi-conductor systems to more full-featured processor-controlled systems that employed smaller skinny (two- or three-pair) wire. Newer systems required less labor and less expensive cabling to install.

Executone was slow to change, perhaps due to an inventory of the older technology systems. Instead of developing their own version, Executone resorted to purchasing from other manufacturers and importers such as TIE (Telephone Interconnect Equipment). In the process, Executone would "private label" the TIE systems by placing their logo on the telephone covers and add a markup before selling the TIE systems to its dealers. Other than a cosmetic change, the same telephone products were also available to their dealers' competitors. Executone dealers soon found that their competitors were buying identical key telephone systems from the same sources for less than what Executone was charging them. The only difference was the shell or cover on the telephones; features and capacities were the same.

Executone also decided to enter the PBX field. The first product they introduced was the E1000. However, they failed to provide adequate technical support and training for their dealers. When problems surfaced, some dealers and their customers became angry and threatened to sue Executone. Brody and Hellering conveyed to their board of directors that they had a serious problem on their hands, and no resolution was apparent.

Meanwhile, after his termination from Arcata in 1972, Dorsey and two other smaller investors purchased Voycall, the company he had unsuccessfully encouraged Arcata to purchase years earlier. After restructuring the company and boosting up the backlog, they turned around and sold Voycall to Arcata for much more than what Arcata would have paid when Dorsey first proposed the acquisition.

Now, looking for a new enterprise Dorsey had been observing a startup company in San Rafael only a few miles from his home. Digital Telephone Systems (DTS) had been developing a digital PBX with a stylish console used for answering incoming calls by an attendant. Don Greene was an engineer who had been selling a few switches to independent telephone companies for their central offices. His new PBX was based on the same design. Dorsey tried to convince Greene to allow him to distribute the DTS PBX to Interconnect companies through a new company, American International Telephone Corp. (AITC), he'd formed with a minority partner, Bob Knutson. Greene didn't like the idea, as his preferred market was the independent telecoms that didn't much care for the Interconnect companies installing systems in their territories. Finally, Greene relented and entered into a five-year distribution agreement with AITC, giving them five percent commissions on all North American sales, excluding public utilities.

Greene, who had not been successful selling the DTS PBX to any independent telephone company, was unprepared and surprised when, less than a week after the agreement with AITC was signed, AITC came in with their first order. Greene wanted to know who the customer was to make sure the sale didn't violate the exclusion in their agreement. Uncertain as to what Greene had in mind, they reluctantly informed Greene that the customer was SP Communications (later SPRINT), a division of Southern Pacific Railway. SP bought a few more DTS PBXs. Green had his attorney confirm that SP Communications wasn't a utility that would have violated their agreement. Gradually, he warmed up and concluded the Interconnect market could be good for the company. Executone would be AITC's next target.

Knutson, through an internal contact, had become aware of the frustration Executone management was facing with their dealers over their current PBX product. Dorsey and Knutson introduced them to the DTS product. It took months and several visits before Executone, Inc. signed an agreement with AITC to distribute the DTS (D1000) PBX through their dealership network, also paying AITC a fee for each DTS sold. While the D1000 began to generate Executone PBX sales with some dealers, Executone had a major problem. Executone had no proactive program to train their dealers in basic telephony or in the selling methods necessary to effectively compete with local telephone companies. Complicating any training was the fact the tariffs that governed the rates charged for subscriber equipment differed for each Bell operating company in the various Executone dealers' territories.

After contracting with AITC for the distribution of the DTS PBX, Knutson visited several dealers demonstrating the DTS and explaining how to sell it. Executone approached Knutson to work for them directly, conducting basic training in telephony and sales methods to achieve successful results. Knutson deferred. He had other plans. Executone began to replace some of their dealers, and other dealers became motivated to sell their distributorships. Dorsey and Knutson bought the dealership in Seattle in 1976, and with Dorsey's long time IBM friend, Jim Healy, bought the Executone dealership in San Diego the following year. Demonstrating significant improvement in sales volume in these two locations, in 1983, Executone also gave Dorsey the dealership in Phoenix after the prior distributorship had been cancelled. Executone began to achieve more success as they replaced older distributorship owners with individuals who had previous experience in the Interconnect industry.

CHAPTER 13

Litton Business Telephone Systems (LBTS)

B y the early '70s, Litton, a publicly traded company, had established itself as a leader in a number of different fields with strengths in government contracting, radio communications, engineering, and developing technical products. What caused Litton to jump into the Interconnect business isn't known, but the timing was right. Litton Business Telephone Systems (LBTS) was incorporated in 1971 and headquartered in Southern California. John Barbour had sold his company, National Communications Planning Service (NCPS), to Arcata Communications in 1969. After leaving Arcata, Barbour was recruited by Len Mende of LBTS. Barbour recalled Len Mende had an Interconnect company in New York and, along with his friend, "Mickey" Weiner, had approached some major companies they believed had the ability to manufacture phone systems and provide them with exclusive distribution. Litton certainly had the qualifications. LBTS developed a sophisticated business plan with the intent to develop their own products after establishing a nationwide network. Regardless of their motivation to enter the market, Litton Business Telephone Systems had a very short history in the Interconnect business—only three years, leaving the business in 1974.

Meanwhile, Arcata National decided to sell Arcata Communications to General Dynamics, who also owned Stromberg-Carlson, a manufacturer of phone systems including PBXs. Arcata Communication was folded into Stromberg-Carlson and moved the headquarters from the Bay Area to St. Louis, appointing former GD executive Don Hyatt as President. The earlier shakeup of Arcata Communication's top management and subsequent move by General Dynamics after the purchase and move to St. Louis created a vacuum for a number of salespeople and managers on the West Coast with experience in the business. A good number of these salespeople were hired by LBTS. Barbour, for example, was a creative salesman and had hired some very talented salespeople—and a few unethical ones.

A culture soon developed in LBTS: "Get the sale at any cost." Former LBTS salespeople spoke about LBTS phone system proposals that included a "finders-fee." The "finders-fee" could be made available to any customer, consultant, or third party that might influence the contract. This form of "bid-rigging" came to light in November of 1973, when four Litton executives were indicted for paying bribes to an employee of San Mateo State College to influence winning of a telephone contract for the college. Within a year and a half after they entered the business, Litton claimed they had twenty-five percent of the Interconnect market. That claim seems to be a bit of a stretch. While LBTS sales on the West Coast were strong, they had virtually no presence in the Midwest, Texas, or Florida where TSI operated. After the indictments, Litton closed down the business.

Even so, during Litton's "wind-down" period, Jack Hart, who was hired to manage the San Diego LBTS office, and Steve Lewis, who ran the Orange County office—both former Arcata salesmen— were still selling about twenty or more systems a month out of their respective offices. Most Interconnect locations at the time were satisfied with three or four system sales a month. Litton's Senior VP, James Mellor, interviewed LBTS executives and salespeople involved in the San Mateo College bid scheme after the indictments were handed down. In a memorandum of Mellor's interview with Len Mende, he told Mende that he had "screwed up a very promising business activity." What makes Litton's story interesting is not what they accomplished in the short period or what they failed to do correctly in the market, but what they did next.

Litton had entered the business in 1971, and elected to close it in 1974. With LBTS offices now shuttered, in 1976, Litton filed an antitrust suit against AT&T. Litton sought $570 million in "lost profits," claiming AT&T used monopolistic, anticompetitive, and predatory practices. The case consisted of four and a half years of pre-trial proceedings, five months of trial, 18,000 pages of testimony, and 945 exhibits. Litton claimed they were "forced out of the market in 1974 due to the PCAs" (the requirement for and cost of AT&T's connecting arrangements). As mentioned earlier, when the FCC mandated AT&T open its network to competition, AT&T argued hard for network protection in the form of Protective Connecting Arrangements rather than establishing a process of certification of non-AT&T equipment. Commissioner Johnson dissented from the order, concluding, "The use of PCAs was much as if an electric utility prohibited customers from using a toaster unless it was designed, manufactured, and installed by the utility itself."

The FCC went along with AT&T's proposed solution of providing PCAs. Litton commissioned and entered into evidence a damage study on Litton's "Potential Loss of Profits" by Richard Hexter, former board member of Arcata National. At the time of the study, Hexter was with Donaldson, Lufkin & Jenrette. The study estimated Litton's market share for the period of 1972 to 1990, and projected that by 1978, AT&T would still have seventy-nine percent of the terminal equipment market and that competition would never exceed fourteen and a half percent of the market. Included in the suit was the loss of $53 million Litton BTS incurred before leaving the market.

In one segment of Litton's discovery process that focused on PCAs, AT&T had faced the prospect of proposing interconnection standards on short notice. In 1967, AT&T had formed a Tariff Review Group—perhaps in anticipation of the *Carterfone* ruling—to review possible tariff modifications. AT&T's Tariff Review Group thought standards were feasible. But they couldn't tolerate the thought that there would be no tariff provisions at all to limit equipment interconnection. Late in 1968, the FCC permitted AT&T's proposed PCA tariffs to take effect. A national Academy of Sciences ten-month report, commissioned by the FCC, concluded PCAs were an appropriate solution that AT&T used to support their position.

Litton also put into evidence a number of AT&T documents to support the contention that AT&T simply could not demonstrate that the PCA requirement was necessary to protect the network from harm. Litton pointed to an in-house report prepared in 1971 by one of two AT&T representatives to the PBX Advisory Committee titled: "A Credibility Gap Exists." The document cited the existence of long periods of time for interconnection on the message (AT&T) network and greater interconnection on private line facilities. The carriers (operating companies) still found it virtually impossible to cite cases of harm. Lack of any evidence of harm would cause manufacturers, users, and regulatory bodies to challenge the expansive efforts that AT&T insisted must be taken to avoid network pollution. A 1972 AT&T report to management stated that there had been no evidence of harm, concluding that PCAs and their rental tariffs would likely be rescinded. Litton surfaced AT&T documents that indicated many at AT&T thought certification would be developed in a matter of months. Further, many at AT&T believed certification was inevitable regardless of their adopted position using protective devices. Nevertheless, in September of 1973, AT&T Chairman John deButts, in a speech before the National Association of Regulatory Utility Commissioners (NARUC) declared AT&T would oppose certification standards because the nationwide switching network was "too valuable a resource to risk a perhaps irreversible threat to its performance that would ensure from fragmentation of responsibility for that performance."

Litton's case against AT&T relied heavily on the fact that they had never been able to make a case for the PCA requirement. In one case, two Atlanta hotels were using the very same brand of non-Bell PBX equipment. No PCA interface was required for the system that AT&T had purchased from a third-party manufacturer and leased to one hotel, while PCA interfaces were required when the other hotel purchased its equipment directly from the third-party manufacturer. An AT&T internal "Tactics Memorandum" stated that, if it withdrew from the PBX Advisory Committee, it would accelerate an FCC decision in favor of certification. Litton claimed the memorandum supported their claim that AT&T did what it could do to delay and obfuscate plans to develop certification standards. AT&T's strong opposition to certification standards was undertaken in bad faith was a principal special finding of the jury on which the verdict against AT&T turned.

In addition to protesting the necessity of PCAs, Litton struck at the delays AT&T used to deploy them. Litton Produced a 1970 memorandum by an AT&T VP stating, "we have been able to placate the situation with explanations of '... possible misunderstandings or only a temporary delay' and assurances [that] no serious supply exists, but doubtful this approach will continue to avoid formal action by the FCC." PCA shortages continued, as evidenced by complaints from Bell Operating Companies. A June 1972 Telex from Illinois Bell Tel to AT&T headquarters read: "so many defective KS2071 [PCA] couplers that we have difficulty providing interface devices and meeting due dates." A letter of October 1972, from an Ohio Bell executive to an Ohio Bell VP, stated: "An increase of a number of vendors have complained bitterly because of our failure to supply [PCAs] ... in many cases we have been unable to quote any realistic delivery date." And in the same letter: "How can we continue to insist on use of [PCAs] when we are unable to provide such a device ... [this] must be given the highest level of attention at Ohio Bell, Western Electric, and AT&T before we end up with a large-scale customer revolt and potential legal action for restraint of trade." AT&T's New England Telephone and Telegraph had an average of twenty-four days to fill requests for PCAs but missed installation dates by an average of ten days. Shortages of PCAs were discussed at a June 14, 1973, meeting of Bell Interconnecting Equipment Coordination Committee. Litton also cited an internal AT&T film simulating an antitrust trial of a suit similar to the one eventually filed by Litton. In the AT&T film, employees were urged to destroy incriminating company documents.

While Litton's case focused on the unnecessary use of PCAs, their unreliability and lack of availability, Litton also provided evidence tending to show AT&T engaged in slash and burn techniques calculated to make cutover from AT&T to Litton equipment as bothersome as possible for Litton and its customers alike. Litton also cited similar experiences other Interconnect competitors faced, like chopping off existing wiring, refusing to acknowledge receiving communication about cutover dates, changing those dates, and failing to provide necessary equipment. The fact that others of us in business were experiencing the same thing in different parts of the country during the early 1970s points to AT&T's coordinated effort to throttle competition.

AT&T's defense of Litton's claims was formidable. They argued that Litton caused its own demise and that PCAs had little impact on their ability to conduct business. In November of 1973, four Litton BTS executives were indicted for paying bribes to an employee of the state college system in San Mateo, California. As part of an internal investigation of possible employee misconduct, Litton attorney, Norman Roberts, made notes of interviews with various Litton BTS employees during the course of his investigation. AT&T argued that Roberts notes constituted a "devastating admission" against Litton as they revealed that Litton employees gave potential customers "calculators, girls, and anything else" to make a sale, and that "skimming" and "funny deals" were commonplace. The trial court permitted AT&T to prove that the officials were indicted and subsequently discharged but excluded evidence of bribery under a Federal Rule of Evidence because of its emotional and prejudicial content. At one point, though, the trial court indicated it would consider admitting AT&T's bribery evidence against Litton's *if AT&T would allow Litton to offer proof that AT&T had bribed public officials.* AT&T declined the offer.

One person I interviewed was the VP of Marketing and Sales for a software company. He wished to remain anonymous relaying an interesting story about AT&T's discretionary use of bribes. The VP's firm had recently been acquired by an AT&T operating company. After the acquisition, a Bell executive was appointed to their board and was in attendance while management was reviewing pending sales. The software company had one large sale pending with a nonprofit organization worth several hundred thousand dollars. They had been unable to close the contract for several weeks. The Bell executive excused himself from the meeting saying he was going to make a phone call. Several minutes later he rejoined the meeting saying nothing. Two days later the president of the nonprofit company called to say they had decided to sign the contract. Upon picking up the contract, the VP was told how appreciative they were of Bell's generous donation made earlier that week. The VP said, "He made quite clear that the Bell donation had made the sale." One can wonder what additional evidence Litton had.

CHAPTER 14

The U.S. Antitrust Case
Against AT&T Builds

The new chairman of AT&T was a fighter. He was not going to let the
Bell Operating Companies and AT&T lose revenue by allowing competition.
That was the spirit the presidents of the Bell Companies were longing for.

I t was raining in Seattle on September 20th, 1973. A little more than a year had passed since John deButts, new chairman of AT&T, addressed his top executives of the various Bell Operating companies in Key Largo and listened to their wails about competition stripping away their lucrative business customers. At that time, he had promised them a comprehensive policy decision about how AT&T would respond to the threat of the "cream skimmers" such as MCI and the various Interconnect companies. Now it was time for deButts to deliver. He chose Seattle, as the annual convention of the National Association of Regulatory Commissioners (NARUC) was held there. The group of state regulatory commissioners was more sympathetic to AT&T's position than the FCC had been. The state public utility commissions lacked the staff the FCC had in determining complex rate structures and frequent tariff increase requests by the local operating companies. These new interconnections allowed by the FCC were causing them additional stress, and many commissioners believed this new competition would disrupt an effective system of telephone regulation, which had been in place for more than fifty years. DeButts had been invited to address the more than 1,000 state utility commissioners, AT&T personnel, and reporters on the last day of the convention. With his North Carolina drawl and powerful voice, deButts was about to make the most costly speech in AT&T history ... and his career.

DeButts opened by speaking about the principles that had guided management of the Bell System for decades: "an unusual obligation [on management] to see to it that the service shall at all times be adequate, dependable and satisfactory." DeButts claimed this obligation had been met, and the only sound policy that would continue to fulfill this obligation was to "continue to furnish the best possible telephone service at the lowest cost consistent with financial safety." He then stated that this policy had been threatened by competitors who wanted "a piece of the action—with no concern for the public at large." DeButts said, "Competition cannot help, but in the long run [will] hurt most people." The Chairman of AT&T had chosen this moment to launch a political campaign promoting the right of the world's biggest company to own a telephone monopoly.

The speech had fallen on an appreciative audience—except for one. Sitting in the audience was Bernard Strassburg, chief of the FCC's Common Carrier Bureau. The Common Carrier Bureau is the regulatory division of the FCC that had recommended the 1968 *Carterfone* decision. Usually, the FCC commissioners went along with bureau's recommendations and did so allowing both MCI and telephone equipment providers to compete with AT&T. For decades, up to the early 1960s, the FCC and AT&T had enjoyed a friendly relationship. Long distance and telephone equipment prices were established through staff negotiations and compromise. If the FCC felt AT&T's requested tariff increases were too high, they would work things out face to face. This negotiation process began to break down during the 1960s. The FCC began to question how

AT&T was going to adapt with rapid technological advances in fields such as satellites and microwave systems. AT&T's service crisis in New York City and other major cities created suspicions in Congress that the FCC was not doing its job. DeButts' comments at the NARUC convention were a direct slight to the FCC and its recent decisions on competition. Thereafter, Strassburg put an end to the informal negotiations with AT&T, and the bureau launched a wide-ranging investigation into AT&T's conduct.

While the Justice Department's Antitrust Division was conducting investigations on AT&T's practices during the early 1970s, the North American Telephone Association (NATA)—the industry association of Interconnect companies, telephone system importers, telephone manufacturers, and telephone supply companies—sought input from its members on AT&T's anticompetitive activities through its attorney, Ed Spievak. Specifically, NATA requested written instances of AT&T's "pre-emptive marketing" (proposing future products not yet tariffed), sabotage, collusion, "un-hooking" (inducing customers to cancel contracts for private telephone systems), and "bad faith" (failing to respond to or deliver protective arrangements for subscribers). Our small firm, TSI, sent in twenty-nine documented cases which we were informed were part of those funneled to Senator Philip Hart, chairman of the Senate Subcommittee on Antitrust and Monopoly, and Philip Verveer, staff attorney for the Justice Department's Antitrust Division. Although relatively new to the Antitrust Division, Verveer had been reviewing AT&T's history and hearing MCI complaints from Larry Harris, MCI's attorney, about AT&T's maneuvers to contain competition. In April of 1974, Verveer drafted an antitrust lawsuit and sent it on to his superiors.

The timing of his draft favored AT&T, as the attorney general, Elliot Richardson, had recently resigned over Nixon's firing of

Watergate Special Prosecutor Archibald Cox. Shortly before his resignation, President Nixon appointed William Saxbe as attorney general. Saxbe was an older former Ohio senator, and few in the Antitrust Division thought him likely to attack AT&T. Saxbe didn't have a reputation for working hard but was known for enjoying his rounds of golf and hunting sojourns. With Saxbe's appointment, Phil Verveer thought the AT&T antitrust lawsuit as dead on arrival. However, Saxbe was searching for a way to reverse public opinion about the Justice Department and internally reestablish integrity and independence. In the fall of 1974, the proposed AT&T antitrust suit reached Saxbe's desk. It was just the right vehicle to accomplish his goal. On November 20, 1974, Saxbe met with several senior antitrust division attorneys. The intent of the meeting was to discuss the AT&T case and to prepare him for a second meeting later that day with AT&T's general counsel, Mark Garlinghouse. AT&T's legal staff had been making presentations to the staff in the Antitrust Division, hoping to persuade them that breaking up AT&T would not be in the best interest of the public. In that second meeting, Saxbe announced that he intended to bring an antitrust action against AT&T.

A hearing on the case was scheduled for February 20, 1975, before Judge Joseph C. Waddy. Waddy had been assigned the government case but had not yet met the attorneys from either side. George L. Saunders, Jr., a partner in the Chicago law firm of Sidley & Austin, had been tapped to defend AT&T's position in the case. Saunders argued that the case was an ill-advised attempt by young Antitrust Division attorneys who wanted to break up the Bell System. He further argued that AT&T was regulated by the FCC and that AT&T only followed what the FCC stated AT&T could and could not do and. As such, it was not a proper case for antitrust laws and didn't belong in Waddy's court but with the FCC. Waddy was swayed and decided the discovery process should be delayed until the jurisdiction question was resolved. The delay of discovery efforts stopped the case cold.

DeButts sensed the clouds of destruction were gathering around him and that he needed a more forceful and powerful strategy. The public must become involved. Newspaper headlines, thanks to AT&T lobbying, were already crying that the pending antitrust suit would unnecessarily break up a good thing and drive up costs for everyday consumers. DeButts concluded the FCC had started the problem in the first place. Instead of returning to the FCC with a plea that competition was becoming uncontrollable, creating havoc in the phone industry, and suggesting that the two should work out the problem together, deButts chose to use might for his cause. With a million employees, a strong labor union (CWA), and an in-place network of lobbyists, deButts could solve AT&T's competition problems through Congress.

THE "BELL BILL:" AT&T'S FINAL BARRIER TO COMPETITION

"We have decided the time has come to call the public's attention to its stake in the matter," deButts said. "Were the telephone companies deprived of . . . revenues from their more discretionary services, they would face the necessity of increasing the average customer's bill for basic service as much as seventy-five percent." A Bill was drafted that deButts called the Consumer Communications Reform Act (CCRA) of 1976. Most everyone called it "the Bell Bill." It proposed that long-distance services of all kinds would become "utility" functions provided by a single, integrated system—which was already built. AT&T's network should be selected as the country's official monopoly. MCI and new long distance companies would be forced out of business. To avoid the certainty of lawsuits and lengthy court battles, AT&T wanted immunization against all antitrust lawsuits and wanted the ability to buy up any of its competitors. The regulation of telephones and telephone equipment would be decided by each individual state PUC and not the FCC—essentially wiping out the *Carterfone* decision. As such, some states might allow the sale of private telephone systems while others could deny competition. Competition in general would be strangled.

An army of lobbyists descended upon Congress and committee members alike. These were not the usual, well-dressed executives, but included Bell System midmanagers and technicians, each flooding their congressmen's offices with a copy of the Bell Bill in hand. AT&T had employees in all congressional districts and made sure they each visited their congressman. The lobbyists would spend weeks following congressmen around and meeting with their staff. The tactic seemed to be working. AT&T and their local operating companies with thousands of voters and a political war chest at its disposal found more than 200 congressmen to sign on as co-sponsors of the bill—although few really understood its ramifications. Vulnerable were the newer, less secure congressmen who grasped what supporting the bill might mean for them. But getting sponsors is a lot easier than getting votes. AT&T spent millions on the lobbying effort, but despite the lobbyist infestation and deButts' constant rhetoric of "competition is not in the public interest," the Bell Bill began to run into trouble.

Representative Tim Wirth, Democrat from Colorado and majority member of the House Communications Subcommittee, had heard complaints from the Interconnect industry about Bell's anticompetitive tactics and had appeared receptive to containing AT&T's unfair practices. A hearing was scheduled to allow AT&T to respond. The room was more crowded than usual. While an AT&T executive was testifying, Wirth asked the executive to identify any of his colleagues who were in the room. After several introductions, Wirth asked, in desperation, "Will everyone associated with AT&T just stand up?" Every person in the room stood, about 150, laughing nervously.

AT&T was having difficulty finding anyone on the Communications Subcommittee to sponsor the Bell Bill. It was eventually submitted by Teno Roncalio, a Wyoming Democrat. Even Roncalio was at a loss to explain why AT&T had chosen him, a lone congressman from Wyoming, to carry their cause. The Bill was losing ground, and the chairmen of both the House and Senate Communications subcommittees were opposed to the bill and hoped to smother it by inaction. Hearings on the Bill opened on September 28, 1976. The subcommittee's chairman, Lionel Van Deerlin, had called the first hearing. Those who were cosponsoring the Bill and were putting pressure on the subcommittee to demonstrate something was happening as a result of all AT&T's lobbying efforts influenced motivation behind the hearing. Congressman Wirth began, narrowing his focus, not on the viability of competition in the industry nor the perceived value of a ubiquitous monopoly, but on how much AT&T had spent on lobbying for the Bill since early 1976. DeButts replied, grossly understating the amount, saying it was about $600,000. Wirth countered by telling deButts that their committee's internal estimate was more like $100,000 a day. Wirth pressed deButts for the number of people engaged by AT&T who were working full-time on lobbying for the Bill. He replied, "I have no idea, sir, but it is very few—very few. Full time? I doubt there is a single person in the Bell System that is working full time on this legislation." The hearing room erupted in laughter.

AT&T's ownership and control of Western Electric had always been an irritating factor within the Justice Department. In the back of the minds of some at the FCC and many at the Justice Department was the nagging question: "Would the public be better served by having the terminal equipment market served by free market competitors?" The early antitrust efforts with AT&T resulted in the divesture of AT&T's investments in foreign manufacturing plants. AT&T also had denied vigorously that it was cross-subsidizing any of its competitive products or services with its non-competitive services. But AT&T was not persuasive. The Justice Department, Congress, and the FCC all contested AT&T's view and worried about subsidies flowing in the opposite direction. As the Justice Department's antitrust suit was bubbling away in the background, William Sharwell, responsible for AT&T's antitrust research organization, responded to this sense of impending crisis by forming a group of AT&T sixth- and seventh-level managers to formulate some new approaches. In the hope that the FCC would be a better bet than either Congress or the courts, Sharwell proposed that AT&T offer the Commission a truce in the field of terminal equipment. By abandoning the terminal equipment market, AT&T would position itself better in the intercity and long-lines market and nurture a more general spirit of cooperation with the FCC. Alfred Partoll, a member of AT&T's legal staff, drafted a letter of reconciliation to the FCC presumably signed by John deButts which alluded to the foregoing with AT&T's acceptance of terminal equipment market and intercity services. The letter was never signed or sent to the FCC.

DeButts had made a major error. In his efforts to lobby the "Bell Bill" he neglected a major political step. AT&T had clearly blundered by not greasing the tracks at the subcommittee level before it tried to move the bill through congress. Had AT&T taken the time to educate at the subcommittee level, their actions would have appeared less heavy handed.

DeButts strong-arm tactics, arrogance, and blustery manner had worked against him. His credibility was shot. AT&T had yet to prove any harm to their network resulting from connection with non-Western Electric products. AT&T argued that the antitrust case should be dismissed as the FCC had jurisdiction. In early 1978, the United States Supreme Court ended AT&T's effort to have the antitrust case thrown out. The court decided that the FCC could not effectively regulate AT&T, and the only proper authority to resolve the competition issue was the antitrust court. Had deButts attempted some concessions with the FCC on competition and had employed fewer intimidating tactics, the outcome may have been different.

CHAPTER 15

TSI Establishes a Foothold

I t was a tough struggle to get the first few phone systems sold and installed. We made twenty-two presentations before landing the first TSI sale. Southwestern Bell's delays with PCAs and promises to potential customers of future telephone products that would be tariffed with reduced monthly rentals made sales more difficult. Midwesterners are generally leery of something new—"conservative" doesn't aptly describe their attitude. The St. Louis business community had a reluctance to change that was stronger than anything I had experienced in other cities. On three separate occasions, prospective customers asked what school I had attended or who my father was. Having never been asked these questions before, I concluded the customer was probably just trying to get comfortable doing business with us. When any of our supplier's sales representatives came to call, such as I did with Steve Sherman from TIE, I would take them on sales calls hoping potential customers would think there was more than just one person in the company.

Revising our sales approach to a consulting one that included first inviting the telephone company to identify any problems, make any recommendations, and propose alternate telephone systems before we made our presentation greatly improved our success ratio. With the client's permission we also contacted their customers to learn how they interfaced with our client—sometimes calling into the client's offices from outside lines to experience what their customers did. It was a more detailed approach with emphasis on the client's customer support. Cautious clients who claimed their customer support couldn't be improved needed to be convinced. I would make several telephone calls to the client's office posing as a potential customer and recording the responses. I would play the recordings back for the client who would listen to hear callers being on hold for long durations, calls extended from the attendant ringing with no answer, misdirected calls, and calls answered by the wrong party without being placed back on hold. Annoyed but interested, the client was then receptive to a TSI presentation. Today's telephone systems cure these problems, but in the early '70s, the problems were rampant.

Although more time consuming, the consulting method again bore early fruit. We began to expand the company once we had about ten happy customers, and moved from the answering service to a 1,500 square foot office/warehouse. To be clear, the reception area was only large enough for a desk and two chairs. My office was not as large. My wife, now several months pregnant, offered to be my receptionist. She had prior experience as a receptionist and PBX operator with Utah Mining and Construction in San Francisco. Although an unpaid TSI employee (as was I), her career didn't last long for three reasons. You guessed correctly on the first one—she was due in a month. The other was that she had a habit of leaning over in her chair so she could see me at my desk, and would frequently ask with a smile: "Shouldn't you be making prospect calls?" Or "Why aren't you out on appointments?" Although always delightful, I had married a micro-manager. The decision to fire her came after an incident with my best customer. After speaking to my receptionist wife, he suggested that I find a replacement. Lynn Pruett owned a large John Hancock Insurance agency in St. Louis. After several weeks of hesitation and a great deal of caution, he finally signed TSI's contract for a thirty-plus station phone system which we installed. It was our largest installation to date, and I relied heavily on his recommendations. Pruett called the office about a minor service item while I was out. My wife started to take a message and began to giggle over his name. She couldn't spell it. He asked her to spell it back to him and repeat the number. Her giggles turned to waves of laughter. Calling the office, she relayed the service message to me. When I met Pruett, the first thing he said was, "If you're trying to build a company, I suggest you replace your secretary, 'Susie.'" I never told him that I was married to my only employee.

While businesses had a natural reluctance to change from the Bell System—their only source of phone equipment for years—we were surprised and delighted by the way our new customers were enthusiastic to help us. Even during contract signing, we asked for a reference, and again, after installation of the system. This wasn't a new sales tactic, but I had never experienced the zeal of customers to help us grow as we did in St. Louis. It was not uncommon for our customers to pick up their handsets from their new telephone and call a business associate, banker, or golf partner and recommend that they meet with us. On occasion, a new appointment was made on the spot.

We encountered one sales competitor more often than any other: Guy Knox, the sales manager from Independent Telephone Systems—the same company whose truck I saw shortly arriving in St. Louis. I called Knox and arranged a meeting. Knox had a tough exterior and was most comfortable with a cigarette dangling from his lips and a beer after work. He was a hard worker, a diamond in the rough, and had the marks of a good manager. We hired him as our sales manager and quickly hired our first two salesmen, Mike West and Tracey Powell, who both later went on to become successes on their own. We began to build a sales team who would later refer to Knox and myself as the "hammer" and "glove," respectively.

We had a short, profitable year in 1972. By the end of 1973, we were approaching $1 million in sales and moved again into a 15,000 square foot building in Ladue, in the west suburb of St. Louis. Also, during the year, we formed Leasetel, Inc. to help finance the phone systems we installed, and were one of the first independent companies to do so. Most customers preferred to lease, as they had previously expensed their Southwestern Bell Systems. Leasing through the company added an additional comfort level to our customers and provided us with a competitive advantage. Toward the end of the year, we also assumed a small customer base of Litton Business Telephone Systems. Shortly thereafter, I was contacted by Gary Witt, former sales manager of Litton BTS. He made a proposal to open a TSI subsidiary in Kansas City and was willing to make a small investment in it. He also brought with him a probable sales manager and a productive Litton saleswoman. As the sales manager had a dubious reputation in California, we agreed to accept the sales manager only on a temporary basis and under a consulting contract that kept him at arm's length.

In June of 1974, TSI was approached by a manager from ITT's Communication and Equipment Systems Division offering to sell their customer bases in St. Louis, Chicago, and Elkhart. This provided an opportunity to expand our customer base and enter the Chicago market. Contained in ITT's PBX customer base were some large hospitals that provided us a platform to pursue other large customers. Guy Knox was promoted and moved to Chicago as branch manager. He asked to bring with him our first two hired salesmen, West and Powell, each later becoming sales managers in Chicago. We had added TeleResources TR32 to our existing sales of ITT, Northern Telecom, Plessey, and OKI PBX products. The

TR32 was our most feature-rich digitally-based PBX product under 200 lines (telephone stations) and contributed our greatest gross margin. We also enjoyed an exclusive distributorship with TR in the areas we then served. During the first few years we enjoyed a verbal exclusive distributorship for key systems (small telephone systems under sixty stations) from TIE's president, Tom Kelly. As soon as we became successful selling TIE, they began appointing other local competitors. We began selling Toshiba and Iwatsu key systems. It was going to be a chase for exclusivity for some time to come. An exclusive agreement from a manufacturer on distribution was critical in maintaining decent margins. We were satisfied with our product line but needed larger full-featured PBX systems to complement our consulting approach. That was about to change.

Stan Blau was president of Consolidated Communications in New York and was a founding member of the North American Telephone Association. He also helped elect me to NATA's Board and was instrumental in our obtaining distribution of the Plessey and TeleResouces products for our firm. In November of 1974, at a NATA convention in San Francisco, Blau introduced me to Ken Oshman and Dick Moley from ROLM who had been observing this new Interconnect industry. At first, I couldn't think of any logical business connection with a name like "Rome." Once the spelling was clarified, the two extended an invitation to me and our Operations VP, Dave Stewart, who was with me at the time. They wanted to show us what ROLM had been developing in their facility in Cupertino, near San Jose. A lot of small companies were introducing devices like auto dialers and answering devices, and we had only modest interest in traveling down the peninsula even though I grew up in the area. Oshman and Moley didn't provide detail about what they were developing, but convinced us that we would not be disappointed. The fact that Blau recommended them was not to be ignored.

We had a pretty busy schedule but agreed to visit them at the close of the NATA convention. As we drove down, I began to recognize the area. The ROLM facility was located in an industrial park that once contained farms and orchards. One farm, where at fifteen I once strung tomato wire and picked cucumber, was now part of a beautiful new industrial park. Everything about ROLM spoke quality. We were impressed with the presentation, the story of ROLM's founding, and the enthusiasm of its people. The demonstration of their computerized branch exchange (CBX) was eye-opening. Other telephone manufacturers were coming on the market with digital PBXs, but this blew everything else away. The features the system offered, its capacity, and the software applications ROLM was developing fit our consulting approach for larger customers perfectly. At the end of the meeting, I was asked to review ROLM's "Preliminary Distribution Agreement." Even though field trials of the CBX had yet to begin, I returned the document to our hosts with my preliminary signature. We were eager for our first installation. When ROLM first began to ship the CBX, our St. Louis ROLM customer, Transamerica Insurance Group, wanted the system installed in their Denver office. UBS in Denver was a ROLM distributor and had not sold any CBXs as yet. This was going to be their first ROLM customer, for which they were grateful. ROLM asked that we just transfer the entire sales contract over to UBS, including all profit and sales commission. Fortunately, we were able to work out a satisfactory agreement with UBS, but it was the first glimpse we saw of ROLM's inflexibility on a distributor's exclusive territory.

TSI's business was doing well in all three locations. The competition in Chicago had well-established customer bases by the time we opened our operations there. Our technicians in St. Louis belonged to IBEW (International Brotherhood of Electrical Workers). In Chicago there were two types of IBEW contracts—one had a higher wage scale than the other based upon the risk and difficulty of work. All telephone cabling our company installed was low-voltage and low-risk, and most of our installations were new construction which was much easier. One would think a lower wage scale contract would be appropriate. John Presti of AllCom, a successful local competitor, warned me that the IBEW business agent was seeking a contract with TSI that paid the higher scale. He also advised that I might need to lose $10,000 in a round of golf with an IBEW business agent to get the contract. Fortunately, the golf game was never proposed, and the sum never paid. Unfortunately, Presti was right. They only offered the higher scale contract to TSI.

The larger Chicago competitors had established sweetheart arrangements with local telecommunication consultants who, mainly for financial reasons, would always recommend them over other companies. We knew how to play that game. We reached an agreement with my friend Darrel Nelsen of Natcom in New York. Nelsen had been the inspiration for starting my own company. We would represent his consulting company, Natcom, in Chicago and St. Louis and, in turn, act as Natcom's agent in the cities we served. Each of us paid our own expenses. I hired two consultants and rented an office on North Michigan Avenue. The operation was structured to stand on its own. The consultants were paid a salary and, upon obtaining a consulting agreement, would establish the specification for the desired system and request bids from Illinois Bell Telephone Company (IBT) and competitors. Natcom consultants would make comparative presentations to their clients which included a telephone system from TSI. Even though the TSI relationship was disclosed to the client, the consultants were biased. The TSI-proposed system was usually not the lowest priced alternative but was most effective in addressing the customer's issues of controlling long distance costs and improving customer service and support.

Though productive, it was a short-lived arrangement. After receiving about a half dozen contracts, I received a phone call from Illinois Bell's competition manager requesting a meeting. He had been in IBT's marketing department and gained a reputation in retaining their customers in competitive situations. He was tall, well dressed, and more knowledgeable than most Bell reps we encountered. Our Chicago office referred to him as IBT's "Pink Panther," as we often found him in competitive situations, snooping around our installed customers locations, and at industry conventions. We agreed to meet at a hotel lobby, a neutral location. The Panther did not indicate that he would be accompanied by an IBT attorney, nor did I anticipate one.

After introducing the attorney, the Panther opened with: "You've achieved considerable success in the first year since you started your operations in Chicago."

"Thanks, the competition here is pretty intense."

"We think you should consider ceasing some of your operations here."

"Obviously, you'll explain why you've reached that conclusion," I replied.

Without commenting, the Pink Panther reached into his briefcase, pulled out a few items, and slid them across the coffee table between us. The items included copies of a rental agreement signed by me for the Natcom office space on North Michigan Avenue and cancelled TSI checks for rent payment. Also included were copies of checks TSI had paid Illinois Bell for Natcom's telephone business lines.

The Panther continued: "We have more. In situations where we find Natcom involved with an IBT customer, we will feel compelled to share with him these documents."

IBT's blackmail tactics came as a surprise. I could play in the dirt with competitors who paid kickbacks to consultants without having to employ the same tactic, but IBT showed they were not afraid of getting their own knuckles dirty either. When I asked them to put their proposition in writing, they told me no. No surprise here. The meeting was over. I assured them they would receive a response the following day. I didn't need much motivation. We already had an issue between the Chicago Natcom manager and a potential customer where they hadn't disclosed that TSI was an agent for Natcom and when the customer found out it cost us a sale. When I'd tried to reach the Chicago Natcom manager by pager a few days before the IBT meeting, he had thrown the pager out the window claiming it was driving him nuts.

Natcom in Chicago was disbanded, the manager fired, and the remaining consultant was hired to work for TSI as a sales consultant. We moved to larger offices in Bensenville, a Chicago suburb. Several months later, we acquired a small independent consulting company, Telecommunications Systems Management (TSM), owned by Bill Gerken. TSM was focused on telephone system design and network management. It became a subsidiary of TSI, and this was clearly stated on business cards and stationery. IBT had no further issues on our consulting business. Consulting provided our best method of exploiting the features of the ROLM CBX. Independent consultants even began to recommend us—one of which had previously only recommended one of our competitors in Chicago. Our policy with any consultant was: We don't pay kickbacks, and we would only pay a consultant for specific work that TSI would normally undertake with full disclosure to the customer and on the condition that none of TSI's sales consultants had previously made a proposal to the customer.

Toward the end of 1975, TSI was featured in a new industry magazine, *Interconnect Journal*. The article was very favorable about our becoming the largest Interconnect company in the Midwest, and reprints were included in most of our proposals.

In the early part of 1976, I was asked to accompany a salesman about to make a presentation to a potential customer. At the end of the presentation, the customer was reluctant to sign the lease agreement and expressed concern about changing from the Bell System to a company in which he had no information on their financial stability—a frequent response from potential customers. The salesman dug into his briefcase and pulled out a current TSI audited financial statement.

"Where did you get that?" I asked the salesman.

"I borrowed it from our sales manager who always uses it."

He handed it to the customer who reviewed it. Satisfied, he proceeded to sign our contract. Returning to the office, I inquired with the sales manager why any of our sales people were carrying TSI's internal financial statements. All the telephone manufacturers we represented provided us with "support letters" that assured the customer of continuing service on their system in the event the local distributor would be unable to do so. The sales manager stated that more than fifty percent of the time potential St. Louis customers would request financial information on our company. Checking with our Chicago and Kansas City offices confirmed requests for TSI's financial information, but with less frequency. Nevertheless, the demand was there. We made the decision to become public with our financial information. We created a marketing piece resembling an annual financial report, which listed our offices and testimonies from some of our Fortune 500 customers, and filled the back page with the names of several hundred TSI customers. The brochure included financial summaries for each year and our first nine months of 1976. We were on track to hit $6 million in sales for FY 1976—a fifty-three percent increase over the prior year. The brochure was an instant success, and requests for our audited financial information ceased. With the exception of Coradian, a New York ROLM distributor who later issued a public offering, other independent regional Interconnect companies were not in the habit of providing financial information to their prospective customers. The financial sales brochure provided our salespeople with an additional advantage over the competition. In today's explosion of the internet and cellular industries, a $6 million annual sales figure is pretty puny. But this was 1976, and it was a new industry with very few success stories.

During the year, Witt decided he and his new wife wanted to move from Kansas City back to Dallas where both were raised. We agreed to open an office there. They would not have the benefit of selling the ROLM CBX or the Northern Telecom SL-1, as Fisk Electric was the authorized distributor for Texas for both suppliers. TSI, Dallas, could sell the TR32 and OKI PBXs as well as the key systems we sold in the other cities. We decided to concentrate on the smaller-sized-system businesses in the beginning.

TSI had been profitable every quarter since our first year, and all growth had been internally funded. Borrowed funds were needed only once in our first year to cover the weekly payroll. The bank required our home as security for the small $3,500 loan and sent an appraiser out to estimate its value. This left my wife with the impression we had to sell the house. We repaid the loan within two weeks. Thereafter we were able to fund TSI's growth through internally generated profits.

As we entered 1977, we expanded the Dallas office into Houston and felt we were on a roll. Although we were experiencing reliability problems with our initial ROLM CBX systems, especially in non-air-conditioned environments in the Midwest where the humidity was high during hot summer months, it was still the go-to PBX for most large customers as its applications addressed most customer needs more efficiently than any other PBX on the market. Early ROLM documentation did not specifically require air conditioning for the CBX. At several locations where air condition was not supplied in the equipment room, technicians had to remove the cabinet doors and place fans on the floor to cool the system. In a few locations during hot summers, our technicians had to sleep on a cot next to the CBX. In the event it crashed during the night, the installer could have it back up before the employees arrived at work. Because the early CBXs were sensitive to high temperatures and high humidity, TSI increased its warranty reserve on ROLM sales which reduced our profit margin.

ROLM worked to support the start of a ROLM customer/user group (RUG) to help address their technical problems and define future features. No other telecommunications manufacturer had taken a progressive step like this before. Because of our common interests, several of the ROLM distributors also formed their own informal group. ROLM wasn't very happy with this. We would meet on occasion to discuss our issues and our possible future with ROLM which had since issued a successful public offering and its stock was now trading at a high of thirty times their earnings ratio. The initial ROLM distributors concluded ROLM had done limited field trials. The reason may have been that the CBX architecture was similar to ROLM's ruggedized mil-spec computer which operated reliably in very hostile environments. Regardless, the distributors were confident ROLM would get a handle on the problems and the system would become reliable—which it did. Also, ROLM distributors agreed we had never seen a telephone system supplier do so many things correctly and innovatively. Before it started shipping systems to distributors, ROLM even shipped the CBX coast-to-coast to test how it made the trip. Their customer support was unequaled. In the long run, we all concluded that ROLM was the horse we wanted to ride to the finish line. There were other topics the ROLM distributor group addressed: Bell System's anticompetitive activities and other systems and products we may consider distributing. We became a close-knit group and socialized together. The group inquired with ROLM senior executives about the possibility of buying its distributors. The answer was they would only consider the purchase of a distributor after "scrubbing" the distributor's financial statements and, if interested, would only offer net book value in ROLM stock. We all felt becoming part of ROLM would be good decision for our respective companies. Our counter suggestions to ROLM for buying its distributors of "one times our annual sales," or four to five times book value considerations, fell on deaf ears.

One problem plaguing all of ROLM distributors was that we were losing margins in the sale of smaller size key systems due to the proliferation of competitors handling the same products. We all wanted an exclusive product that addressed the small business market. The alternative was to ignore that segment and concentrate on the larger systems—which would have pleased ROLM and larger system manufacturers, but had an inherent risk of dependence on one or two manufacturers. Another factor was that the best way to teach professional sales consultants to recognize and define the needs of the larger customer was to start them on small systems first. Many salespeople washed out in the first six months of their training. If they were successful in selling smaller systems first, they became more confident, experienced in understanding telephony, and less dependent on the sales manager to help make the sale. With success in smaller systems, their success in larger sales was enhanced. None of the ROLM distributor group wanted to abandon the smaller market.

Stan Blau, who had introduced several of us to ROLM, met with our group. He had identified two possible Israeli companies that had the technical capability and willingness to help define and manufacture a system we wanted and provide us with exclusivity. With the help of Blau, a group of ROLM distributors including TSI and Blau formed Pentacom, Inc. Each of the distributors would be able to sell the product on an exclusive basis in their areas, while Blau would appoint noncompeting distributors in other areas. Two of us flew to Israel for a week to work with Telrad on defining the product and describing desired features. I returned two additional times before prototypes were sent to us. The product was the "Key BX." Targeted at the small business market, it supported up to sixteen business lines and thirty-two telephone locations, expandable to sixty-four—totally modular with many features not yet on the market.

For the larger PBX portion of the market, TSI's sales of ROLM and TeleResources were increasing, while Northern Telecom SL-1 sales were slipping due to a lack of features that benefited large customers. Our Northern Telecom district manager, Carl Bagwell, was understanding, but TSI risked losing exclusivity on the SL-1. When Bagwell visited our offices in St. Louis toward the end of 1977, he conveyed a piece of information that was to change our direction.

Winter Park Telephone Company, an independent telephone utility based in Winter Park, Florida, had established an Interconnect subsidiary, First Communications with branch offices in seven cities in the state. First Communications sold various key systems and the Northern Telecom SL-1, and was an exclusive distributor for ROLM in Florida. The company had finished the year at $2.2 million in annual sales. Bagwell thought Winter Park Tel was interested in divesting of First Communications, as it didn't complement their long-range objectives. They had created an entity that was feeding off their own established base of utility customers— their bread and butter. He also thought Northern Telecom was attempting to buy the First subsidiary. We jumped at the opportunity and flew a small team to Florida to meet with Winter Park execs. We completed the sale on January 10, 1978, and issued a press release the following Monday.

Bagwell, representing Northern Telecom, had worked extensively with the personnel at First supporting First efforts in sales and installations. He had always been professional and demonstrated management skills in working with our salespeople. All the First Communications knew Bagwell and respected him. We appointed him Vice President and General Manager of First Communications. It was an easy transition. A few days after the announcement, I received a less than friendly call from ROLM's president, Ken Oshman. We would have thought ROLM would have been thrilled with our purchase of First, as we just protected ROLM's interest in Florida. Had Northern Telecom bought First Communications, they would have cancelled First's ROLM distributor agreement. That was not the way Oshman saw it. I was reprimanded for not having communicated with ROLM prior to the acquisition. Oshman informed me that ROLM had been considering purchasing First and that TSI had pre-empted them. In reflection, perhaps I should have conferred with them. ROLM might have said, "back-off," or suggested a joint venture in Florida. But I wasn't used to conferring with others regarding decisions on the growth of our company, and ROLM hadn't done a joint venture with any of its distributors—yet.

1977 turned out to be a banner year for TSI. We finished the year with $8.3 million in sales—a forty percent increase over the prior year, earning a pretax profit of slightly over $1 million. First's financials were not included for the fiscal year, however the cost of the acquisition was. We condensed the offices in Florida down to five, giving us a total of nine operating centers and two offices for TSM, the consulting company we acquired, and Leasetel. We decided to issue a press release on our earnings. As we entered 1978, we published a professionally produced twenty-one page audited annual report that also featured ROLM and described some of their applications that provided first-time solutions for many of our customers. It became a great sales tool and attracted some unexpected attention.

We received an additional boost when the editor of *Telephony*, a long-standing industry publication widely read by employees of Bell and independent telephone companies, visited our offices and conducted an interview about TSI. The cover story on our success appeared in the March 27, 1978, issue. TSI received additional national coverage that fall. A UPI reporter asked for an interview. Thinking some local exposure would help, we granted the interview. During the last week in September of 1978, the article was picked up by forty-two papers across the nation, including the *Chicago Tribune*, several Florida papers, and the *San Jose Mercury News*, in my old hometown. The UPI article helped sales in cities where we had offices when the article appeared.

The author after reporting first $1 million in profit.

A few years before, we had been approached by Diversified Industries, a St. Louis company that made railcar wheels. Diversified offered $1 million for the company, then increased the offer to $2 million. Even though our sales were about $3 million at the time, I declined. After our 1977 financial statements were released, our consultant, Len Saxon, arranged for a Saturday morning meeting with the newly elected President of Chromalloy Industries, a St.

Louis based publicly traded conglomerate. The president was young, dynamic, and unassuming—perhaps in his mid 40s and quick to get to the point. We had already sold a 500-line ROLM system to one of their subsidiaries, CPI. We hit it off immediately and quickly reached an agreement. Chromalloy would purchase TSI for $7.5 million—mostly cash. The offer was penciled out as an informal term sheet with a formal document to follow, subject to Chromalloy's board approval. He took with him the term sheet with both our signatures. My wife, Susie, was delighted with the news. Maybe her husband would be able to have more family time now. On Monday, Saxon called to tell me the young president died of a heart attack over the weekend. At Chromalloy, an internal struggle began for his replacement, and we never heard about the offer again.

Toward the conclusion of our scheduled Pentacom meeting, copies of TSI's annual report were distributed to each member. A colleague paid a compliment on the report and added that "Most everything you've done has become profitable." While flattered, the comment made me feel uneasy. Things had been going very well for us. All tides rise and fall. As 1978 provided more national exposure for the company, the year would provide opportunities to make several mistakes.

TSI produced an internal financial statement within ten to fifteen days after the close of each month. On the last day of the month, each branch reported labor, material costs, work in process, etc. Contracts were immediately sent to St. Louis, and a job progress file was opened. In the early part of 1978, our accounting staff experienced difficulty in obtaining operating information in a timely fashion from a few of the branches. Don Danner had been an auditor with the accounting firm Massie, Fudemberg, Goldberg & Co which the company used since our inception. Danner was young, intelligent, and enthusiastic. After our first full year, we hired him to head up our accounting department. He was an excellent fit, and our auditors supported the move. To compensate for the increase in business and delay in collecting financial information, the accounting department began to "common size" the financial information from the delinquent branches. This meant the reporting of income and costs was based upon prior average costs and profitability for the branch, planning to adjust any discrepancies in the following month. We were losing control. When reviewing our staff and structure there were very few college graduates on the payroll. Most critical, none of us had prior management experience in a company of TSI's size. We needed more depth.

I tasked Saxon again to find us a financial VP. A former ITT executive, Jim Van Cleave, was located and interviewed. He was hired as TSI's Vice President of Finance with a promise, tied to favorable results, of becoming Executive Vice President. Van Cleave had been on Harold Geneen's staff. Geneen, President of ITT, had a reputation of insisting on quarterly reporting and each division meeting quarterly objectives. Apparently, ITT division presidents would meet with ITT's corporate staff and review the prior quarter and forecast for the next quarter. If the division or subsidiary failed to meet objectives, they were given warning and required to define how the CEO planned to correct their path to achieve the objectives. After a few quarters of failing to meet objectives, the staff would meet again the following quarter with the division president or CEO. With them, they would bring a replacement for the division in the event the objectives had not been met. A consistent motto of Geneen's was: "Words are words, explanations are explanations, promises are promises—but only performance is reality." Van Cleave reflected that attitude. I should have been more sensitive in handling his transition into TSI. Danner immediately submitted his resignation—a loss I regret. Van Cleave didn't waste time. He began to tighten up TSI's financial reporting, replacing Danner with a new financial manager. Believing the company lacked a strong foundation for growth, he added key functions to TSI, hiring a general counsel and marketing manager, and strengthened contract administration and management information systems with the help of a new Wang computer. We now had a "Corporate Staff," and the group moved into a building next door.

ROLM had given notice to some of its distributors that they were planning to establish a new distributor in Michigan and invited distributors who were interested to submit a business plan. We put together a plan and used the opportunity to introduce Van Cleave to the ROLM executives. When we presented the plan at ROLM's offices, we received a polite but tepid response. They inquired about Van Cleave's ITT background but didn't focus much on the business plan. Other than Fisk Telephone Systems, a division of a large electrical contracting company in Texas, we believed TSI held the strongest financial position. We didn't think Fisk was interested in expanding into Michigan, whereas we had customers in neighboring Illinois. ROLM told us they would get back to us. At the end of the meeting, they stressed they preferred to see us put more emphasis in growing the ROLM base in the areas TSI was the authorized ROLM distributor. Perhaps they were still festering about our acquisition of First in Florida. My friend, Marty Liebowitz from Electronic Engineering (EECO) in Ohio, was going to present his plan for Michigan the following week. ROLM decided to accept EECO's plan and never provided us with their reasoning—probably thinking they didn't need to. When ROLM's decision was announced, it came with a surprise. ROLM did a joint venture with EECO for the Michigan distributorship. We didn't know if that was EECO's initial strategy or if ROLM suggested it during EECO's presentation. TSI was having growing pains, though, and it was probably best that ROLM decided to reject our business plan.

Working sixty to eighty hours a week for the past six years was taking its toll. My family didn't see much of me during the week, and at best I contributed a couple of Sundays during the month. Before Van Cleave arrived at TSI, my stress level had become difficult for others around me to endure. I was unaware at the time that some employees entering my office would first inquire with a secretary if I was "wired," meaning that I might become explosive over an issue. After Guy Knox was hired, he introduced me to his neighbor, Dave Stewart, a graduate of Washington University in St. Louis and managed a computer staff there. Although this work had little to do with telephony, he had more knowledge about computers than anyone I had met. His wonderful, up-beat, unassuming attitude made him likable to everyone—like a big lovable bear. As telephone systems were becoming computer driven, hiring Stewart seemed like an opportunity not to be missed. Stewart became Vice President of operations, working diligently with manufacturers, technical issues, inventory management, and labor unions. We paid to send him to ROLM for technical training on two occasions. ROLM charged all distributors for training technical personnel and required that technicians carry with them a complete ROLM spare parts kit. For years, I never had a problem with Stewart, but I managed to make a critical management error. We had been having some technical issues with a ROLM customer in St. Louis that we had not been able to resolve who was threatening to throw his ROLM system out and go back to Bell. I had asked Stewart to go meet with the customer, address their issues, and engage ROLM's assistance if needed. Seeing Stewart two days later in the lunchroom, I inquired how the meeting went. He hadn't met with the customer. Without thinking where we were and with others present, I berated him for not following through. Big mistake. I should have suggested we move to my office or apologized on the spot. The damage was done. A week later, the marketing manager for OKI, one of our suppliers, called. Stewart was in their offices with some other TSI employees announcing they had formed a new company and were requesting a distribution agreement from OKI. I've always regretted my error. It cost the company several valuable employees.

During the last part of 1977, we decided to host a telecom conference at a Chicago Hotel. We invited over 250 local telecom managers, corporate executives, consultants, hospitals, state agencies, and current TSI customers. The response to the invitations was more than eighty percent and filled the room. ROLM was the featured system, and several ROLM executives came in to make presentations. Other telecom conferences would follow, but ours may have been one of the first that featured a private telephone system—at least in the greater Chicago area. The response was more than what we could have hoped for, and several sales followed shortly thereafter. Toward the end of the conference, a TSI saleswoman happened to be at the pay phones at the hotel and overheard a ROLM manager next to her calling back to ROLM headquarters stating: "In spite of themselves [TSI], the event was successful and should result in several sales. However, all of us here have received comments from many potential customers about how they would prefer to buy directly from ROLM rather than a distributor."

TSI salespeople also heard these comments from potential customers throughout our history, so I didn't think much of what she related at the time. ROLM had always assisted us by providing previously mentioned "support letters" to customers, as well as accepting phone calls from potential customers and answering their questions. With ROLM's support, we had been able to secure contracts with several fortune 100 companies, such as IBM, Xerox, Sears, Transamerica Insurance, ITT Aetna Financial, Owens Corning, as well as four locations of Illinois Tool Works, the Federal Reserve Bank, universities, and hospitals. TSI's credibility had been steadily growing with large companies, and we felt confident that we could assuage any concerns.

At the conclusion of TSI's telecom conference, Knox and I met with two ROLM executives who began to pressure us to concentrate more on selling ROLM CBX systems rather than other products we distributed. I started to explain, again, how ROLM didn't fit all applications. In the smaller sized market, ROLM's rich features would not be utilized, and the CBX wasn't cost effective. Before finishing my comment, Knox, who didn't like the pressure, jumped in, "When you guys make the same kind of margins we do, then you can tell us what to do—until then, FUCK YOU!"

While ROLM's revenue was larger than TSI's at the time, our gross margin as a percentage of revenue was higher. An awkward silence followed, and a disagreeable look flooded the faces of two ROLM executives. Even though the Chicago office had endured a lot of CBX problems and failures during the year, it wasn't in TSI's best interest to antagonize our major supplier. Always professional, neither ROLM executive said anything. The meeting was over.

Despite these growing pains, we were optimistic about the future and eager to add First Communications' sales and profits to TSI. However, something was taking place at the end of the year outside of TSI that would seriously impact the company. We would not learn about it until a month later. A positive event occurring at the end of 1978 that would improve our business as well as all companies competing with AT&T in future years. The unnecessary and dreaded PCAs mandated by AT&T would be eliminated.

CHAPTER 16

TIE Communications (Telephone Interconnect Equipment): A New Leader in Small Business Key Systems

After his tour with the Navy was over in 1961, Thomas L. Kelly used his Navy training in communications to secure a job with the New York Telephone Company (New York Tel or NYTel). Up to the early '60s, it was common practice for NYTel to embed some of their major account reps in their larger customer's headquarters. Paid by NYTel, major account reps were provided an office and would help Fortune 100 corporations manage their telecommunication needs. However, no such benefit was provided for the midsized corporations and institutions. In the early '60s, NYTel decided to discontinue the practice. Several of the large corporations began to hire former NYTel employees to replace the perk they once enjoyed. After a few years of experience in auditing NYTel customer inventory records, and familiar with Bell System tariffs and equipment, Kelly and fellow colleague, Ron McLeod, resigned NYTel and started their own consulting company, Professional Telephone Consultants (PTC), leasing space at 404 Park Ave in 1967. At the time, Kelly was living in Manhattan and McLeod in the

Bronx. They knew that midsized companies and some of the larger firms who had not hired a telecom professional would likely be ripe for consultants who could do the job and spare the corporation the expense and commitment of hiring a full-time employee. They were right. PTC became a very good business. Most of the time, they would work with a customer for a continuing flat fee based upon the work required. On few occasions, PTC would work on a contingency basis, taking a portion of the savings they would achieve through implementation of PTC's recommendations. After work, they would frequently enjoy a cocktail at the Biltmore Cafe across from their office. FCC's decision on the *Carterfone* case in 1969 was about to change Kelly's direction—unwillingly.

Gil Engles first met Kelly and McLeod when the two worked for New York Telephone. Engles had been a top salesman for Stromberg-Carlson since 1958, reporting directly to its CEO and Chairman, Dause L. Bibby. Some of Engles' customers, such as the transit authorities in New York and New Jersey, were also clients of PTCs. To shorten his commute at the end of the week, Engles first would return to NYC from Stromberg-Carlson's home offices in Rochester, New York, before proceeding on to New Jersey where he lived. While in NYC, he was known to have a drink with Kelly and McLeod. Prior to the FCC's *Carterfone* decision, Stromberg-Carlson only sold equipment to independent telephone companies. At the time, telephone manufacturers were terrified of selling to anyone other than operating (regulated utilities) companies and NYTel wasn't buying telephone products from Stromberg-Carlson—only Western Electric.

As Kelly recalls: "One afternoon late in 1968 or early 1969, Engles was having a drink with us at the Biltmore. We were discussing the business and the changes that might occur as a result of the recent Carterfone decision. PTC had a client, J&S Plumbing, located in

Queens, that had been using an old manual cord board PBX. As the PBX offered no internal dialing, the firm used a paging system to contact employees in their warehouse. We had recommended that they replace the old PBX with a new Stromberg-Carlson PABX that included a built-in intercom and a cordless console instead of the unsightly cord switchboard. Engles agreed to sell us the SC PABX allowing PTC to install it for our client. After the client accepted our recommendation, we had to implement the system installation in two stages. NYTel refused to connect the SC PABX to their network and blamed the delay on the fact that Western Electric had not provided them with the line interfaces [PCAs]. After we filed a Public Service complaint in Albany, NYTel allowed us to connect the SC PABX directly to their network with the understanding that, when the interfaces were ready, we would connect [the system] to their devices. So, we 'hard-wired' the system to the NYTel trunks. That was the first Interconnected system in New York. After that installation, Engles gave us two additional leads which led to the sale and installation of Stromberg-Carlson Cross Reed PBXs."

A New York investment firm and client of PTC, D.H. Blair & Co., had made an investment in a startup company, MEC in Greenwich, Connecticut. MEC was importing a small key telephone system manufactured by Meisei in Japan. A principal at D.H. Blair asked Kelly if he would go up to Connecticut and take a look at the startup. Kelly responded, "Why would I want to do that?" PTC's consulting work was expanding, and he didn't want to take time away from his core business.

"We'll make it worth your while," came the reply.

Kelly went. After seeing the Meisei key system, he thought, "Gee, this product might be good for our smaller clients." Kelly returned to D.H. Blair and conveyed his thoughts: The company needed an engineer who was familiar with U.S. telecommunication networks, and Kelly concluded the company president was a problem. Blair tasked Kelly with finding an engineer for the startup. Kelly knew of one—Steve Kerman. Kerman lived in the Bronx and was very familiar with almost all aspects in telephony. In fact, in his small apartment, Kerman designed and built his own central office. Kelly flew to Japan and informed Meisei that he was sending over an engineer for two months to work with them. Meisei wasn't happy with the arrangement but went along with it. Kerman was supposed to return at the end of two months while staying in contact with Kelly in New York. Kerman stayed at Meisei working on the product for a year. On his off hours, he tinkered with the hotel PBX where he was staying. Finally, a prototype system was sent back to the states for evaluation.

D.H. Blair had decided they wanted Kelly to run the startup, MEC. Kelly didn't want to take the offer. PTC was doing well, and he didn't want to have to move to Connecticut. Besides, as he told the investment firm, he felt the president of the startup was a concern, lacking the ability to grow the company. Shortly thereafter, Kelly's contact at D.H. Blair removed the president and came back to Kelly again with an offer. Kelly talked it over with Gil Engles and finally decided, "Why not?" Kelly told D.H. Blair that he was willing to "play around with the startup for a while but would not commit to more than one year."

The company changed its name from MEC to Telephone Interconnect Equipment, and incorporated in Greenwich, CT, in 1971. The startup had been working out of a converted plumbing house that had also dispensed propane gas. It was agreed that McLeod would continue to run PTC, which was doing well, since there was no certainty on how the new startup was going to fare. From the start, Kelly had decided he wanted out of the arrangement and brought in Gil Engles to organize the company and head up the sales effort. Kelly was intent on importing a product that didn't require any cosmetic changes—thinking that it would be accepted by U.S. dealers for competitive features which Western Electric systems lacked. Kelly soon realized that Meisei was a reluctant suitor. "Meisei didn't want to do this," Kelly said. While working with Meisei, Kelly's agent in Japan introduced him to the chairman of Nitsuko. Both Meisei and Nitsuko were part of the Sumotomo family of companies referred to as a Zaibatsu. Prior to WWII, the Japanese government encouraged wealthy families to invest in previously government-controlled companies. Looking much like large conglomerates, Zaibatsus were controlled monopolies that were vertically integrated. The Sumotomo Zaibatsu included Nippon Electric (NEC). Kelly decided to change from Meisei to Nitsuko products.

The first product that Kelly showed Dorsey and me, we concluded looked like the head of an alligator, was probably made by Meisei. He demonstrated it in the back of his van at the Arcata Communications sales conference at Glen Cove, Long Island in 1970. We didn't like it. Most dealers didn't like it. The other product Kelly demonstrated was made by Nitsuko. It only came in gray with a rotary dial and buttons or keys next to the dial. While not that great looking, it was better than the alligator model. The Nitsuko phone also came with a lot of features new to the U.S. market at that time: music-on-hold (that was the biggest feature, even though it only played "Annie Laurie" from a music box in the main control unit), privacy, conference, and hands-free voice announce—and, as I recall, one could also use the speakers in the phones to page. It was nicknamed "The Gray Whale." All we knew about Kelly was that he had a van. Regardless, he was given an opportunity to make a presentation at the Glen Cove conference and allowed to sell the gray whale directly to the Arcata branches. TIE management quickly realized that, while the whale—now also in black with a new feature of "hands free answer back" on intercom calls—was quickly replacing AT&T key system equipment, feedback from the distributors was that they still considered the whale unattractive.

Steven Sherman was hired by Engles to become TIE's first salesman. Sherman had also worked for New York Telephone, and had been doing phone audits for Phone Consultants of New York which was owned by Art Gallob. As Sherman remembers Engles: "He was the son of German immigrants, joined the Navy and was thought of as a 'hat and tie professional from the old school.'" Also, both were pretty big drinkers. I didn't drink until I was twenty-six. Engles always acted as a gentleman—at least that was my experience—always courteous and got down to business quickly. Kelly was also friendly and approachable. Both liked to imbibe given an occasion."

One night in Manhattan, Arcata alumni Darrel Nelsen and Bill Dwyer were having dinner with their wives when Kelly stopped by the table and wound up joining them. After several drinks, a tired Kelly fell asleep at the table. A discussion began about who should drive him home. Nelsen and Dwyer decided it would be best to have his wife come get him. She agreed to come, but instead of driving Kelly home, she joined the group. They had a great time into the early morning, all while Kelly slept, head on the table. That was about the same time that Kelly, driving home late one night after a few drinks, missed his driveway and proceeded to drive his new Mercedes into the neighbor's pool.

Kelly was also one of the founders and strong supporter of NATA, our industry association. Kelly states, "TIE took out full page ads in *The New York Times* and the *Washington Post* with the heading, 'CAN YOU AFFORD TO LIVE IN THE U.S. OF AT&T?' The ad described how AT&T owned eighty-seven percent of the nation's prime AAA debt, raising the cost for all borrowers. We ran the ads for six months."

AT&T's Touch-Tone™ service had recently been introduced in a few cities, but the word was out, and businesses were eager for the improvement over the telephone's rotary dial. TIE quickly capitalized on the AT&T technical advancement and developed a rotary dial touch-tone replacement. In most cities the Bell Operating Companies' central offices were not yet updated with receivers that would allow the use of Touch-Tone™ phones. TIE got around the problem by introducing a touchpad that looked more high-tech, although one could hear the key presses convert to out-pulse dialing the same as if a dial on a rotary phone had been used. Through the efforts of Steve Sherman and other regional managers, the company expanded its distribution. By the time Litton Business Telephone Systems was in full swing in late '73 and '74, TIE was the dominant supplier of small business key telephone systems in the U.S. The company changed its position of not making cosmetic changes and began to introduce a plethora of different systems and telephone cases so as to provide for "private labeling" and allow the various distributors and supply houses to have some differentiation.

TIE was also the first to introduce "skinny wire" systems in the '70s. Early key systems allowed the end user to directly access a business line or trunk by depressing the button or key on the telephone associated with that line. With larger PBXs, access to an outside trunk was gained by dialing "9" from the phone, reducing the number of wire pairs from the PBX switch down to just two for a single line telephone. In a "key" telephone system, each telephone station needed enough buttons or keys to accommodate the number of business lines required by the business and required a pair of wires for each line. There were three limiting factors as to the maximum number of telephones in any key system: (1) The size of the telephone could support only so many keys or buttons, and some keys were also used to access features. (2) The more restrictive limitation was the size of the cable or wire—more lines, more pairs of wires. (3) The number of "links" or intercom communication paths in Western Electric key systems had only one or two links—used both for the distribution of incoming calls and internal calls. Before the introduction of processor-controlled systems with four conductor (two-pair) wiring, all key systems made by AT&T's Western Electric, Stromberg-Carlson, ITT, and Northern Telecom, needed either twenty-five- or fifty-pair cables. Capacity was reached at the point where the number of business lines required more buttons that the system could support. In the early to mid '70s, these key systems usually had a capacity of thirty to forty stations. The larger cable also carried a higher price than two-pair wires and required more labor hours in pulling the larger, heavier cable through conduits or mounting on walls. In 1974, at the International Telecom Conference held in Geneva every four years, Gil Engles walked into GTE's booth where Automatic Electric had their key systems on display. He showed them TIE's new two-pair processor-controlled phone. As Engles recounts, "They were blown away. It devastated the industry." TIE had changed the course of the small business telephone market. Others were soon to follow, including Inter-Tel, Toshiba, Iwatsu, and Northern Telecom. Ron McLeod joined TIE after the company moved from offices in Greenwich to Stamford but resigned several months later.

In 1975, TIE received a second cash infusion of $192,000 from the Connecticut State Product Development Corp. After they issued a public offering, the state received back $590,992—or a return of 300% on their investment. By the mid to late '70s, TIE could probably claim fifty percent of sales of all private key system sales sold in the U.S. It seemed like all Interconnect companies sold

TIE, although they might have another product or two. TIE established four regional offices, with Steve Sherman heading up the North East region. Sherman established a close relationship with Engles and Kelly. Engles recalled, "Sherman even came to communion with us."

Steven Sherman (L) presenting TIE achievement award to distributor J. Michael Jarvis of Jarvis Corp.

Sherman's jump to Inter-Tel some time later was a major and bitter disappointment for both Kelly and Engles. His departure notwithstanding, as TIE moved into the late '70s and early '80s, nothing on the horizon appeared to threaten their dominance in the small business market segment. Continued success was inevitable. But just how permanent is success?

CHAPTER 17

ROLM: The Startup that Changed the Industry

A mong the student body at Rice University during the late '50s and early '60s, there were four bright young Texans who excelled in their engineering classes, some graduated at the top of their class. The oldest, Walter Loewenstern, graduated Rice in 1959, then served two years in the Navy. A recruiter, Burt McMurtry, an earlier graduate of Rice working for Sylvania, had spoken to Loewenstern before graduating and suggested he call after serving his military obligation as part of his ROTC program. A major incentive at Sylvania was their work-study program with Stanford. If you qualified for the program at Sylvania, you would receive full pay while taking time off to study at Stanford for a master's degree or doctorate—the tuition for which was fully paid by the company. Loewenstern went to work for Sylvania. Gene Richeson, a Rice 1963 graduate, was also recruited by Sylvania and went to work in the same military defense division as Loewenstern. The Sylvania recruiter worked with Rice's engineering professors to determine which graduating students might make a good fit with his employer. A professor had asked if he had interviewed Ken Oshman, who was top of his class. When McMurtry was able to speak to Oshman, he learned that he had applied to Harvard, intent on earning a master's degree and then starting a company. Oshman's intended pursuit of management plus engineering was an unusual and attractive asset, but McMurtry's push toward a career at Sylvania at the time didn't seem to influence Oshman, who had already applied to Harvard. McMurtry was persistent and wrote Oshman about laser development work at Sylvania and the company's work-study program. The combination must have swayed Oshman, as he also went to work for Sylvania working in laser development.

The following year, McMurtry returned to Rice to recruit the top engineering graduate of 1964, Bob Maxfield, who was also being recruited by IBM. IBM had promised Maxfield that he would have an open hand working with computers, which was of keen interest to him. Maxfield had earlier visited the Bay Area while interviewing at Sylvania and had dinner with Oshman at McMurtry's suggestion. Liking the area, Maxfield accepted IBM's offer with the condition that he could work with computers at IBM's plant on the southern outskirts of San Jose. Maxfield received a fellowship from Stanford and resigned IBM to pursue his doctorate. The four—Richeson, Oshman, Loewenstern, and Maxfield—all earned their doctorate degrees at Stanford.

Loewenstern also had ambitions to start a company and continued to look for opportunities while working for Sylvania, which had since been purchased by GTE (General Telephone Equipment). Loewenstern proposed a new venture to his management: development of a police vehicle tracking system. When management declined to pursue the project, Loewenstern continued to work on the project on his own time. He finally decided it would be best to inform his manager what he was up to. His manager demanded that he stop pursuing the project, which led to his departure from the company. After Loewenstern mentioned to others that he had quit Sylvania to work on a project, Oshman called him suggesting they team up. The two considered several ideas besides vehicle location. They decided to involve Gene Richeson who had been working on military project at Sylvania for five years and was thought to be full of ideas. Richeson held a high respect for his two alumni, especially Oshman's intelligence, and decided to join the group. Of the ideas they considered, most centered on the use of computers. As the three had little experience with computers, they decided to contact Maxfield. When Oshman called him before receiving his doctorate, Maxfield said he had no definite plans. Oshman suggested he join the three Rice grads. He considered Oshman "a natural business genius" and agreed.

The four entrepreneurs defined several markets and products in their business plan that they then presented to venture capitalists in the area. None, initially, expressed interest. Jack Melchor, who had run Sylvania's Electronic Defense Laboratory, founded and successfully sold a couple of companies before establishing a small venture capital fund. After reviewing the group's business plan, Melchor told Oshman that they were trying to do far too much and suggested paring down the plan. After the meeting with Melchor, the four reconvened and began to discard all non-military markets and applications. At the time, computers were being built by major companies, such as NCR, IBM, Sperry, and Sylvania, to meet the specifications of a military program—mostly "one-off" designs with individual programming language. These computers would often take more than a year to develop and deliver. After delivery, the military owned the design and, if more computers for the same application were needed, they could put the design out to bid. These systems were normally of mainframe design and were quite costly. Unless the program language was the same and application very similar, it was not likely to be comparable with the next military-specific program. Mainframe computers were quite large and generated enough heat to require spacious, air-conditioned rooms, with raised floors to accommodate the miles of cabling connecting the computer cabinets.

By 1968, several of the major companies began to introduce smaller computers aimed more at commercial applications of finance, inventory control, and manufacturing—light-duty stuff. The four principals began to focus on the need for a rugged military specific minicomputer that could withstand jolts, harsh vibrations, and inhospitable temperatures—high humidity to below freezing. An added bonus would be if the rugged minicomputer could also run commercially available software programs. The entrepreneurs intended to develop a mil-spec computer that would be in continuous production and available on short notice, rather than a year. They concluded the best way to accomplish these objectives would be to license a commercially available computer and ruggedize it. The question became: "Which computer company should we approach to obtain a license . . . and how do we prevent them from listening to our business plan and going after the market themselves with a similar product?" They figured the larger the computer company— like DEC, IBM, or HP—the bigger the risk. Data General, a smaller startup with only thirteen employees at the time, seemed to be doing things right. The DG Nova minicomputer was both powerful and inexpensive—about $10,000. The four thought, by Maxfield's estimate, they could ruggedize and sell it for about $30,000. There was an inherent risk dealing with a small startup, but their chances of reaching a decision within a short period outweighed the risk.

Oshman and Richeson called the president of DG at home and arranged a meeting in Massachusetts. They came back with a term sheet outlining a licensing agreement which barred them from commercial markets and gave DG five percent equity of their new company. They named the company Datel. The business plan was revised, and Oshman again approached the venture capitalists. The four had no prior experience in the areas of responsibility defined in their business plan, and after a few rejections, Oshman returned to Melchor. During this time, Oshman and Loewenstern were still employed at Sylvania. This time they offered Melchor twenty percent of the company, becoming an equal partner with the rest. After thinking about the proposition, Melchor accepted and agreed to help raise venture capital financing. The five each contributed $15,000 for a total of $75,000. With Melchor's help, they also secured a $100,000 line of credit. Another company had already registered the name Datel. A new name was needed, and the group settled on ROLM—in capital letters, the first letter of each of the four founders' last names.

Starting in an abandoned prune drying shed in June 1969, the four went to work. The shed was in an area of Cupertino that had once been farms and orchards now in the process of being developed into an industrial park. One of the first problems the four encountered was that the circuit boards of the DG Nova were larger than what was needed to meet military specifications and fit into smaller openings. Components had to be fit into smaller layouts on printed circuit boards, and connections had to be able to withstand rugged use. Almost everything including the cabinet had to be re-designed. The milestone that Melchor set was to be able to demonstrate the ROLM ruggedized Nova at the forthcoming Fall Joint Computer Conference in Las Vegas just months away. Hiring a few more engineers and technicians, they were able to have two models at the show—one operational and one non-working that could be viewed through a plexiglas top to show the interior. But the working rugged Nova wasn't actually quite working. At the time Oshman told the story in his understated manner, he just said they had to rely on a hidden cord connected to a DG minicomputer under the counter.

Later, I was told that the problem was due to a delay from the sub-contractor of the memory module. By the attention their booth attracted and later feedback from interested parties, the product was deemed a success, and ROLM had met Melchor's benchmark.

The product had performed well under tests at ROLM, but now had to pass tests for heat, cold, vibration, shock, and humidity by an independent testing lab. It passed, meeting requirements for air, ship, and ground situations. More funds were needed to start manufacturing. Melchor helped raise the next $600,000 through three venture capital groups, including his own. ROLM's first fiscal year ending June 1970 was $240,000 and a loss of $350,000 due to first-year startup costs. After that, ROLM was profitable every quarter, going on to hit a target of $1.5 million the following year. Oshman and Maxfield's roles became solidified, but Loewenstern's and Richeson's became less defined. Richeson was uncomfortable as VP of marketing and sales. Oshman took over the role, and after Richeson bounced around the company in various positions, he left ROLM to return to ESL, his former employer. Loewenstern had been handling administrative duties before going into sales for the company. He found he enjoyed sales, and after taking a sales training course, was approached by Oshman who talked to him about representing ROLM in Washington, D.C. Loewenstern accepted the job of presenting ROLM to military contractors with government contracts where the ROLM mil-spec computer could become the central processing unit.

ROLM continued to develop follow-on mil-spec computers with expanded capabilities, while their market grew into areas outside of the military that required a rugged computer. Revenues for 1973 increased, but profits declined a bit. Oshman's initial estimate of the total market for minicomputer sales to the military was $100,000,000. He was becoming concerned that the total market could limit the size ROLM would grow and began thinking of other markets the company might pursue while complying with the original agreement with Data General.

When ROLM needed a larger telephone system to provide service for its 100 employees, Pacific Bell delivered an antiquated monster. The more Oshman looked into the telecommunications marketplace, the more confident he became that it presented a new growth area for the company. Even though competition in the telecom marketplace had started a few years earlier, the largest U.S. PBX manufacturers—Western Electric, Stromberg-Carlson, and Automatic Electric—were delivering electromechanical switches to their own operating companies. None seemed focused on the use of computers to power their systems. Only a handful of manufacturers at the time were developing digital systems using computers. TeleResources had made inroads with its computer-driven PBX in the small to medium market. ROLM acquired access to one and, taking it apart, concluded that it was cheaply constructed and probably less reliable than one that they could develop themselves. ROLM could capitalize on its experience in the redundancy and ruggedness of their mil-spec computer products. However, a development of PBX was not just a computer employing off-the-shelf software. The PBX would need to be a system solution using proprietary software with specific applications. With acute awareness of what resources ROLM would need to produce a desirable product, Oshman and Maxfield recruited a couple of system engineers from HP. By mid 1973, a team of three began working on the design.

Next, Oshman sought out a marketing manager to complement the team, and recruited Dick Moley, the commercial marketing manager for HP. Moley had a master's degree in computer science from Stanford's work-study program. After arriving at ROLM, he first sought information from large companies to find what they wanted in a telephone system. The research he performed was quite logical—but had probably never been done before by the major PBX manufacturers that supplied systems to utilities. The operating companies were already aware of what their customers wanted. What both large and small companies wanted was not what the telephone operating company intended to provide: the ability to identify and control their telecom costs. Improving service to their customers would just be an added bonus. It seemed that the larger the company was, the higher the percentage of telephone expense was in relation to the company's total sales. Quite often we would find the telephone expense could be more than ten percent of total expenses. Those expenses included the monthly rental of telephone lines or trunks, the telephone system equipment including the instruments, and usage. Controlling the cost of local and long-distance calls was a hot item for all. The cost of making moves, changes, and additions added another layer of expense.

Some companies delayed moving plans because installation and "basic termination" charges on top of the cost of moving the phone system was too high. Bell operating companies concluded that the initial installation charges imposed on large PBX customers did not cover all their costs and would calculate an amount they would need to recapture the "extra" costs should the customer move to a new location in a few years. The customer couldn't influence what these extra costs would likely be and were never provided a calculation of how the operating company would arrive at the basic termination figure. The additional installation charges were amortized over several years—usually ten. Should the customer need to move in a few years after the first installation, they were obligated to pay off the remaining balance.

Moley prepared and presented the business plan to Oshman. Initially, ROLM would sell through distributors on an exclusive basis, and the ROLM CABX (computerized automatic branch exchange), as it was first called, would offer applications that included management's ability to control long distance costs and usage. Oshman loved the plan, but some board members resisted. They claimed that the company's new direction would risk the profitable mil-spec business—the entity that would provide the initial funding for the telecom division. Another concern for some, including a founder, was the risk of trying to compete against the largest company in the U.S. Oshman and Maxfield prevailed, and the project was launched.

Besides appointing regional Interconnect distributors, ROLM also decided to target independent telephone utilities which had more stringent requirements for reliability. The CABX, later reduced to CBX, would be fully redundant, including processors that would switch back and forth so that a failure of one would automatically switch to the other. Remote polling provided the ability to check the health of the system from a service center or secondary location. The initial target market was to address customers requiring eighty to 400 lines (telephones). Smaller customers were less concerned about outgrowing the system. Larger firms, particularly national organizations with in-house telecom managers, insisted on 100% expansion capability in a system. That meant a customer requiring 350 to 400 lines at installation may not be considered unless it could be expanded to 700 or 800 lines. This requirement added an additional complexity to the CBX development. Electronic systems were more susceptible to electrical power line fluctuations than older crossbar mechanical PBXs, and the first CBX systems shipped were found to be extremely vulnerable to power surges and power reductions. Distributors were forced to add expensive Topaz power transformers to stabilize power fluctuations coming from the power source.

As distributors, we had never experienced anything like ROLM and its people. They were the brightest and most responsive we had experienced in the industry—from the founders to recently hired technicians who were dispatched to a customer site to help fix a problem. In our limited experience, only L.M. Ericsson in early 1970s came close in technical support. LME also extended very liberal credit terms to their distributors. But in the mid 1970s, they were still shipping large crossbar systems that did not have the software feature applications of the ROLM CBX. ROLM seemed to be doing everything right.

Any new system that came into the market had birthing problems. ROLM was no exception but was reluctant to admit what the problems were with early shipments. We all had confidence in their ability to resolve "bugs." The technology used in older large PBX products—such as ITT, Northern Telecom, Nippon (NEC), and Ericsson—were less sophisticated, as they had been influenced by AT&T's Western Electric, but they had become reliable. Most of ROLM's early distributors had experience with some of these manufacturers and were eager to introduce ROLM's software features that the market had never before seen. In 1976 and 1977, most non-Western Electric PBX manufacturers offered minor features, such as music on hold, flash-hook hold from a single line telephone, transfer of an outside line, do-not-disturb, and three-party conference. A few offered direct-inward-dial—calls that bypassed the attendant's position and were routed directly to individuals or departments. Not great features, but certainly improvements over Western Electric PBX systems. During the same period, ROLM introduced and offered features that appealed to medium-sized companies and, most importantly, for large national corporations who felt AT&T just wasn't listening to them.

These were cost-savings features customers longed for that no other telephone maker was providing:

- **Least Cost Routing**: The system looked at the digits dialed and, by priority, would place the call on the least expensive path (WATS line, tie trunk, Foreign Exchange Line, etc.).

- **Programmable Ring Back**: Using programmable timers, unanswered calls would ring back to the operator or be routed to another station. The feature was also used to call a user back who tried a specific trunk or station which was busy at the time.

- **Call Detail Reporting**: This management tool provided detail on any call made by any station or group of stations. This included the station placing the call, the time of day, and length of call. Alyn Essman, president of CPI, one of our first ROLM installations, plugged in employee home phone numbers and was "absolutely floored" to find how many were spending hours a day calling home. Two employees would call home in the morning and have someone place the handset next to the TV. They listened to the TV the entire day. Personal phone usage dropped dramatically the following month.

- **Automatic Call Distribution (ACD) System**: A "must have" for any firm with a large volume of calls for sales, customer service, reservations, etc., the software provided an equal distribution of incoming calls to a department or group while providing management with the ability to measure the performance of each agent in the department.

- **Centralized Attendant Service**: Stores with multiple locations in a large area could have all incoming customer calls dialed to any location answered by a single group of attendants. Calls would then be distributed to the desired department in the store nearest the caller. The feature eliminated the need for an attendant at each store and reduced the total number of required attendants.

One can easily appreciate how these features hit the marks of the major companies Moley researched. These features reduced and managed costs while improving responsiveness to incoming callers and providing better service to their customers. Oshman once told me that he considered Moley a marketing genius. It's likely none of the ROLM distributors would disagree with his statement.

There was one limitation to the early ROLM CBXs. Executives who were used to having key systems—multi button phones—in executive suites behind their PBX didn't care for a single-line telephone to perform all the ROLM features. A secretary or assistant might support one or more executives which would share multiple extensions from the PBX along with an intercom linking all with single digit dialing. For the first few years, distributors would install key telephones similar to those the Bell System provided, normally made by ITT. In 1978, ROLM introduced its first electronic telephone station that enhanced use of the CBX features. In the same year, ROLM also introduced both a smaller version, the SCBX for up to 144 stations, and the LCBX, almost doubling the capacity of the CBX. For the next eight years, ROLM would continue to introduce both smaller and larger systems and improve on its suite of features. Distribution was also increased to twenty-three distributors and thirty-seven independent telephone utilities.

In ROLM's mil-spec division, the company had always shared a direct link to the customer. In the telecom division, ROLM was one step removed. Management began to contemplate opening their own direct CBX sales and service offices. Earlier, ROLM had to take over three distributors in Denver, Salt Lake, and Boston who had experienced financial problems. In 1978, they began to systematically open their own ROCOs—ROLM Operating Companies—starting in Chicago, expanding into New York City. Only in a few cases, Virginia and Philadelphia, did ROLM make an acquisition. In some cases, they would wind up competing directly with their former distributor; in other instances, they would elect to not renew the distributor's agreement as they moved into its once-exclusive area.

ROLM continued to grow, its stock price in the '80s increasing twenty-fold from its initial offering. Software features and executive telephones continued to add capability while system capacities also increased. By 1985, ROLM's VLCBX could handle up to 10,000 lines.

CHAPTER 18

Inter-Tel: Persistence, Persistence

One man's motivation to build a company may have come
from a finger poke in the chest by his main supplier.

B orn in 1943, he was the second oldest of five brothers in a family that struggled making ends meet. After fifteen years of marriage, his mother and father divorced when Steve Mihaylo was about nine. His mother was forced to take work as a domestic helper—cleaning other people's homes—as his father contributed very little in the way of child support. Shortly after the divorce, the oldest brother ran away. Unable to keep the brothers well fed and clothed, Mihaylo's mother sent the other four brothers to live in foster homes. These homes were not a happy environment for the brothers, who often went hungry and were severely disciplined for minor infractions. Mihaylo's mother was able to complete school and became a nurse and for a short while, enabling her to bring the four boys back. The arrangement didn't last long, and again she sent the boys to live with different families. After wrecking a friend's truck just after his seventeenth birthday, Mihaylo's father marched him down to the induction center and forced him to enlist before he finished high school.

His experience in the Army helped pull Mihaylo away from rebellion and put him on a new course. After basic training, he was placed in mechanics school at Fort Ord. He then transferred to the 101st Airborne at Fort Campbell where he found an opening at the

Army radio and radar technicians' school. From there, he transferred to the Southeast Signal School in Fort Gordon, Georgia. Mihaylo thrived in Signal School, earning the equivalent of a two-year engineering degree.

Upon his release from the Army, he headed to Phoenix and was hired by Western Electric which had a contract with the government to maintain its underground communications systems that was designed to withstand a nuclear attack. He soon developed a reputation as a problem solver and was promoted to field service engineer. However, Mihaylo felt advancements in Western Electric would be long in coming and decided to get his college degree—majoring in accounting and finance. He enrolled in both Orange Coast College and Cal State Fullerton, completing his course work in two and half years, taking twenty-two to twenty-four credits a semester. College counselors usually recommended their students only take twelve to fourteen. To help put himself through school the first year, he worked the graveyard shift at Douglas Aircraft as a welder. During his second year at college, he applied for a position as installer and service technician at Panoramic Audio, a small company owned by Conway Chester that sold paging, intercom, and background music systems in the Los Angeles area. Chester was impressed with the hardworking young man and allowed him flexible hours to help his coursework.

Upon graduating from Cal State Fullerton, Mihaylo was offered a position with Arthur Young & Co. at a salary of $10,000 a year—normally considered a "plum" opportunity for any new graduate. He delayed starting until the end of summer. Chester, in the meantime, allowed Mihaylo the opportunity to try selling for Panoramic Audio and sent him to a sales training class. When it came to competing against the phone company, Pacific Bell, Chester sought out the advice of Craig Dorsey whose company, CCI, in San Francisco, had developed a sales process that was effective in augmenting phone company equipment with hands-free intercom systems. In turn, Chester taught the same selling method to Mihaylo. Mihaylo was a natural at sales and enrolled in some professional sales courses including one conducted by J. Douglas Edwards. The summer Mihaylo sold systems for Chester, he earned $20,000. Realizing he could make a lot more money in selling, he decided not to pursue a career in accounting.

While at Panoramic Audio, Mihaylo started thinking about working for himself. He met with one of Panoramic's suppliers, Stentafone, a Norwegian based intercom manufacturer, that agreed to supply him with product. However, Mihaylo didn't want to compete directly with someone who had given him his start. Instead, he decided to start a company in Phoenix, where he had worked for Western Electric. Chester convinced Mihaylo that he would need financial assistance and advice. Thus, the new company, Inter-Tel, was formed as a division of Panoramic Audio with Chester and Mihaylo each fifty percent partners. However, it was up to Mihaylo to sink or swim, as Chester decided not to put any cash in the new company. Instead, he helped Mihaylo access product lines, and his wife, Nancy, did Inter-Tel's bookkeeping—for a fee. Initial sales may have been difficult, but Mihaylo was confident in his ability to sell, install, and—with his engineering background in the Signal Corps—maintain the systems he sold. His first two customers wound up being lifelong friends and were constant sources of referrals for Inter-Tel. Mihaylo sold telephone systems to both of them—one a Nippon PBX, the other a key system made by Automatic Electric. The fees for accounting services that Nancy charged were more than what Mihaylo could pay. With relatively few installations in Inter-Tel's base, Mihaylo decided to buy out Chester's fifty-percent interest for $25,000, a good return on Chester's zero-dollar investment.

Knowing that he could sell more systems if they had additional features, he approached his contact at Automatic Electric (AE). On the way to dinner with him, Mihaylo described applications that the new features would address. He knew they were all technically feasible. Mihaylo described his contact as "an arrogant middle aged marketing executive who took little interest in me." While waiting to be seated for dinner, the Automatic Electric executive turned to face Mihaylo, and, poking a finger in his chest, said: "Sonny, if you're so smart, why don't you do it yourself?" Mihaylo decided to do just that. He concluded that the only way to ensure his new company's success and create the features his customers wanted was to have control of his own product.

Frank Woods at the U.S. Department of Commerce gave Mihaylo a list of hundreds of Japanese manufacturers that might be able to supply Inter-Tel with product. He would not be content selling someone else's product; he wanted to have control of the design and distribution. Mihaylo culled the list down to twenty, sent out inquiries, and received back a few responses and brochures. Next, he flew to Japan to begin searching for a manufacturer that was willing to work with him. He decided on Taiko Electric Works who agreed to provide Mihaylo control over the manufacture and design of Inter-Tel products. Taiko already had developed a small key telephone system that he felt was attractive and rich in features with a capacity of thirty telephones. One major condition of their agreement: each order to Taiko had to be worth $2 million. He borrowed and pledged as much as possible and ordered $2 million worth of Taiko phone systems—a warehouse full. He named the product "Key-Lux."

Mihaylo then approached Craig Dorsey. After his role as president of Arcata Communications, Dorsey had formed a new company, AITC. Dorsey and his partner Bob Knutson agreed to distribute the

Key-Lux to dealers nationwide. However, AITC's enthusiasm for the Key-Lux was short lived. In the meantime, TIE, who had been importing key telephone systems from Nitsuko, was offering a new key system with features equal to the Key-Lux, plus one more: "hands-free-answer-back." The feature allowed a called party to respond to an intercom call without the need of lifting the handset. Without the feature, the Key-Lux would be considered "outdated." After a couple of months talking to dealers—many of which considered the Key-Lux to be ugly, Dorsey and Knutson retuned to Inter-Tel and cancelled their distribution agreement. As Mihaylo states: "They basically came back and said, 'we quit, we can't sell this.'" But Mihaylo could—he had to, and did.

Moving into 1973, Mihaylo was facing severe financial problems. About the same time, some telephone manufacturers and a few Bell Operating Companies were introducing Touch-Tone™ telephones instead of the old rotary dials. TIE was quick to seize on the opportunity and introduced a touch-tone pad on their key systems that could be deployed anywhere. The Bell System central offices were not all equipped to accept the unique Touch-Tone™ signals, so TIE included a rotary conversion. TIE customers could claim they had touch tone service. In reality, little time was actually being saved, but the tone pads felt faster and appeared more technically advanced.

It was on Mihaylo's shoulders to now dispose of the Key-Lux inventory without touch tone pads. It would take a lot of perseverance and long hours. The effort to do so consumed Mihaylo for the next eighteen months. Flying his own plane he called on as many as thirty or forty dealers in a week, trying to get back to Phoenix before Friday morning so that he might make a sale Friday and install it before leaving Sunday to see more dealers. Mihaylo also called on Ed Brody, President of Executone, Inc. He agreed to provide Executone a royalty fee for the Inter-Tel products he sold to their dealers. With

Executone's approval, Mihaylo called on them. He would first present the dealers with Taiko's Centurion PBX and would frequently be asked if he had access to any key systems. Using a reverse selling technique, he would pull out the Key Lux, saying, "I'm not supposed to show you this, but..." He would leave with a Key Lux order in hand. In 1974, when Litton exited the Interconnect business, they wanted to sell their inventory of 15,000 touch-tone pads, which normally sold for $15 each. Mihaylo jumped at the chance and purchased the lot for $1 each. To save on shipping cost, Mihaylo shipped them in his luggage, taking three days to get them all to Phoenix. He repeated the process to get them all to Taiko in Japan, saving thousands. Mihaylo hired an engineer experienced in touch-tone receivers who worked with Taiko on the conversion. In the process, streamlined the way the system was engineered, going from a totally wired system to one with PC cards that could easily be added or replaced, ultimately increasing the Key-Lux capacity to sixty stations/telephones.

It took eighteen months to clear out the more than $2 million worth of Key-Lux inventory, but in 1975, it was finally dwindling down thanks to Mihaylo's tenacity. Now it was time to grow the company again. Mihaylo met Steve Sherman at a NATA convention. Sherman, along with Stan Blau, had help setting up distribution networks for Plessey and TeleResources. After leaving TeleResources, Sherman had joined Tom Kelly and Gil Engles at TIE, becoming one of their first salesmen and ultimately director of sales. Sherman helped put TIE on the U.S. map as they grew in the next four years to dominate the small business key system market. Sherman told Mihaylo that he didn't have any stock in TIE or stock options and wanted to increase his income rather than just through sales. The two could not have been more different personalities. Mihaylo was reserved, frugal, and unemotional. He addressed problems calmly and tended to be quiet, listening to others before he spoke. Sherman on the other hand was emotional, peripatetic, explosive, persuasive, and fast talking. Regardless of their differences, the pair appeared to be a good fit. Sherman had the skill and experience to build up Inter-Tel's distribution network. He signed on for an eight percent commission and began calling on his old TIE dealers. Many dealers, including me, found increasing numbers of local competitors selling TIE that Sherman and the rest of the TIE sales force established with each visit to our offices. Saturating the major markets with numerous TIE distributors helped drive their revenue, but it became a frustration with the larger dealers. With several dealers selling the same product, or a TIE product with a few cosmetic changes, Sherman found a ready market for dealers looking for an exclusive product that would help protect their dwindling margins.

Mihaylo envisioned that a combination of direct sales offices and exclusive dealers would be most effective. Sherman was intent on appointing all the dealers he could, much as he had done with TIE. By 1979, however, Inter-Tel only had three direct offices—two of which were in Arizona. By the end of 1975, Inter-Tel had finally broken the $1 million mark in sales. In the late 1970s, the country began to plunge into a severe recession, again placing financial stress on the company. Using the financial understanding he gained in college, Mihaylo kept close watch on receivables and minimized inventory levels. With his large supply of Key-Lux systems having been sold to dealers, he was able to turn inventory more frequently. Sales in 1979 would reach $7.8 million.

As Inter-Tel approached the next decade, their Key-Lux and Key-Lux II consisted of all wired logic and relays at a time when the rest of the competitive industry was developing digitally based microprocessor controlled systems. Inter-Tel dealers were clamoring for the next generation product. Mihaylo decided the best way to create a new system would be to bring Taiko into the development effort. This would reduce the financial burden such a project would require. Taiko agreed and sent two Japanese engineers to join Inter-Tel's team of four. The team was minuscule compared to the hundreds of engineers working at larger telecom manufacturers such as NEC, Northern Tel, ROLM, or AT&T. The team worked through 1978 and 1979, having to develop and create test tools for the new system as test equipment at the time was not commercially available. By 1980, the team had created two prototypes. The new system was labeled the SPK—Stored Program Key system. Inter-Tel would claim it was the first company to offer a microprocessor controlled with special features that could be programmed into the system to meet customer requirements. With a capacity of 120 phones, the SPK also penetrated a market once controlled by PBXs by reaching above the fifty-station market. Sherman, now executive Vice President, acquired eight percent of Inter-Tel for $100,000.

Inter-Tel's sales of $10.3 million for 1980 showed modest growth over the prior year. The following year, they were shipping SPK systems out the door unlike anything the company had ever seen. The dealer network had been hungry for a new advanced product for years. The demand seemed to be insatiable. By the end of 1981, Inter-Tel pushed sales above the $40 million mark. Management at the company was euphoric—everyone was in high spirits. Business publications extolled the virtues of this fast-growing Phoenix-based company. Mihaylo used the enthusiasm and momentum to fulfill another of his objectives. In February 1981, Inter-Tel issued its first public offering. The offer was underwritten by Bache Halsey Stuart and E. F. Hutton and sold out in the first thirty minutes. The initial offer was at $12.50 a share. In a few weeks it doubled, and within a couple of months reached a peak of $54. The following year, Goldman-Sachs and E. F. Hutton did a second offering. Inter-Tel also secured OEM distribution agreements for the SPK with Datapoint, North Supply, and Honeywell Communication Services. The agreements provided them with individual station designs and features. Finally, with a firm financial foundation, several OEM agreements in place, and a network of dealers clamoring for the new SPK, Mihaylo was riding the crest of a successful wave. It seemed like nothing could thwart Inter-Tel's success. Success, however, is sometimes temporary.

CHAPTER 19

Mitel Corp.: The Story of Two Brits

I t is believed the two young men met at MicroSystems International, a chip-making operation and a subsidiary of Northern Telecom. Born in Sussex, Michael Cowpland left England at age twenty-one to work for Bell Northern Research. Working while continuing his education, he received a master's degree and a doctorate in engineering, became a design project leader at Bell, and was the manager of circuit designs for Microsystems International Limited from 1968 to 1973. Terence (Terry) H. Matthews was born in Newport, South Wales, the same year. He was an obsessive fixer of things, from clocks to cars. At age sixteen, he obtained an apprenticeship at British Telecom's research labs. In 1969, he received a bachelor's degree in electronics from Swansea University. Matthews also left Britain and joined Microsystems where he met Cowpland. The two young Brits shared a common interest in working on cars. In their spare time, the two talked about product ideas and starting something together on their own. They decided they wanted to design and make tone-to-pulse converters to be used in central offices based upon Cowpland's Ph.D. thesis. They sought the approval of their employer who provided it based on the premise the two were first going to import electric lawnmowers from England, which was their intent. Revenue from the sale of lawnmowers would provide working capital for their new company. Thus, the name of the company they presented to their employer was Mitel, allegedly an amalgamation of **MI**chael and **TE**rry Lawnmowers. However, Cowpland would later claim the name stood for **MI**ke and **T**erry **El**ectronics. Had they not been given their employer's blessing, ownership of Mitel's intellectual property rights could have been challenged.

Through funds of their own and a bank loan, they raised $4,000, and created the holding company, Wesley Clover, to facilitate Mitel's founding in June 1973. The first transaction the two made consisted of a shipment of electric lawnmowers from England. But the company failed to sell a single lawnmower. It was a fiasco; the shipping company carrying the first batch lost the container. The company initially thought the container had been lost at sea. The lawnmowers were later found, but when they finally arrived, the ground was covered with snow in the Canadian winter, and no one would buy them. Matthews later said, "That taught me a key lesson—the importance of timing. The shipping company lost the lawnmowers! By the time they showed up, no one wanted them, as you can't cut grass when it's covered with snow. I couldn't even give them away."

Although various sources tell pretty much the same story, in an interview with Craig Bailey, he writes that Matthews claimed the story was wrong, and "the shipment was actually lost."

Most of the $4,000 had been spent on the transaction that went bust—little remained. Matthews and Cowpland turned to their original task. The telephone industry was starting to introduce push-button Dual-Tone Multi-Frequency or Touch-Tone™ dial pads. However, most of the operating telephone companies' legacy central offices could not accept the DTMF tones. They needed something that would convert the tones generated into "dial pulses" that the central offices were equipped to receive. "We worked around the clock to develop a tone receiver on a single card," Matthews said. Mitel's cards could be sold at $150 each with an eighty-percent gross margin, compared to big competitors selling their units for $1,500 each. The technology was hot. The duo raised $120,000 in capital from friends and family to make and deliver the first products, and never looked back.

Buoyed with the success of their tone receivers, the duo quickly realized the significance of the then-new microprocessor functionality controlled by software that could be incorporated into the design of PBX systems. Mitel expanded into the semiconductor field by acquiring Siltex, an ISO-CMOS silicon foundry which had gone bankrupt. The following year, in 1977, they introduced their first PBX product, the SX200, with deliveries beginning in 1978. In contrast, ROLM, who started earlier, employed a minicomputer rather than microprocessors in their CBXs. Although the first SX200s were analog and not as full featured as ROLM's CBX, they were priced right and very competitive in the under 100 lines (phones) market. The product was compact, easy to install, and reliable. It was a success.

The two founders were similar and dissimilar. Both were talented, but credit for "marketing flair" was initially given to Matthews, and "engineering genius" was the label placed on Cowpland. Each appeared driven by ambition. In print, Matthews was described as a history buff, dynamic, energetic, animated, ruthless, brash, pugnacious, and demanding, and was once quoted as saying, "I take concrete with my cereal." Matthews' motto was: "Don't be boring, do something," or "make a mark, don't be part of the living dead."

Cowpland was the more physical of the two, barrel chested, with the ram-rod posture of a drill sergeant. An avid tennis player, in his early 50s he reached the men's finals in tennis in the veterans' division at Wimbledon in 1995. Additional descriptors of him included:

wizard, flamboyant, quick to respond to email and voicemail (not a trait often attached to most engineers or "geniuses") enthusiastic, confident, and guarded in his decision-making process. One colleague called him "a man in constant motion." Cowpland gained a reputation for high living, turning a large mansion outside Ottawa into a 19,375 square foot, "grandiose, glittering, glass pleasure palace" in which he stored his sports cars. The large home apparently caused quite a stir in the neighborhood. The rewards for accomplishment vary by individual, including cars, multiple homes, yachts, jets, wine collections, and charities. The material things may have been what motivated the two.

Sales of Mitel's SX 200 ballooned. Realizing the market would demand more, the company made several software and hardware revisions, expanding the SX200's capacity. In 1985, Mitel introduced a hybrid SX-200 digital PBX. Expansion was created by adding additional cabinets, first to 250 lines, up to a 672 port PBX that was introduced in 1978. Matthews was quoted in a *Guardian* interview with David Now saying that Mitel sales "doubled every year for the first eleven years and generally in electronics a successful company can do that and remain stable. Anything above that can be unstable." By 1982, Mitel boasted sales of $250 million and 5,000 employees.

Instability would strike the following year with sales dropping by twenty-five percent, apparently because of the AT&T breakup. However, there was another problem. Mitel had announced that it was bringing to the market an integrated SX 20000 that would grow to several thousand lines. IBM had been seeking an avenue to enter the telecom marketplace, allowing them to augment their data products. Mitel and IBM entered into an informal agreement, with Mitel promising to deliver the product within months. Several industry insiders and some investment bankers doubted that Mitel would be able to deliver. IBM, after several months of seeing no progress, broke off the agreement on June 17, 1983. Mitel's stock dropped twenty-five percent upon the announcement. A few months later, IBM bought fifteen percent of ROLM, eventually acquiring the entire company.

The following year, British Telecom bought controlling interest in Mitel. A short time later, both founders would resign. Cowpland would go on to create Corel, which in 1996, would acquire Novell, the developers of WordPerfect. Matthews would start Newbridge Networks (NbN) in 1986. In 2000, NbN reported revenues of CAD $1.8 billion for 1999, and 6,500 employees. The same year, Matthews sold NbN to Alcatel for a reported CAD $10.32 billion. Matthews then shelled out CAD $230 million to reacquire the network division of Mitel, taking it private and making several investments in enterprise voiceover IP technology (VoIP). At one point, Matthews stated that he had invested over CAD $600 million in Mitel Networks.

Matthews and Cowpland were not timid about starting new companies, buying companies, or investing in them. Though personally making a profit, Mitel struggled to report profits to its shareholders. After Matthews bought back Mitel in 2001, revenues during the five-year period between 2003 and 2007 would hover between $350 million to $380 million a year, incurring losses of between $70 million and $30 million each year during the same period. Mitel, in 2007, purchased Inter-Tel, a U.S. competitor who was moving into the VoIP space.

At the time of the Inter-Tel acquisition, Brian Riggs, Principal Analyst at Enterprise Communications, stated that the merger would require: "[a] fair amount of consolidation due to product lines, markets and technologies that overlapped ... such as the Mitel SX-200 and Inter-Tel Axxess, the Mitel 3000 and Inter-Tel 7000. The combined company will either have to integrate separate communications platforms or make some tough decisions on which products stay and which go. In upcoming months and maybe years, there is no reason the combined company can't continue selling two separate product lines in two largely separate customer bases in separate regions."

Easier said than done, as other major players found out decades earlier. Mitel's CEO at the time, Don Smith, stated, "We believe the merger will deliver value to customer, channel partners, employees and shareholders while making us the clear choice for new clients."

Inter-Tel's revenues at the time matched those of Mitel's. However, Inter-Tel had been consistently profitable, while the new Mitel reported losses for every year between 2001 and 2009—except in 2008 when the company reported profits of $12.6 million. Mitel needed to write off the "good will" from the Inter-Tel acquisition, including $284.5 million in 2009, when the company reported net losses of $193.5 million.

After the Inter-Tel acquisition, Mitel's combined yearly revenues hovered around the $600 million mark between fiscal years 2010 to 2013. The company achieved income during those four years of $37.2 million, $88.1 million, $49.8 million, and $6.2. million, respectively. Revenues between 2014 and 2018 jumped dramatically to $1 billion while net losses began to mount for the same four years. The revenue growth was largely due to a number of acquisitions, including Aastra Technologies (11/2013), Mavenir Systems (3/2015), Polycom (4/2016—but fell apart that November), and ShoreTel (September 2017). By the time of the ShoreTel acquisition, the company had over 4,000 employees.

In 2016, the company reported a loss of $252.3 million from discontinued operations and net loss of $217.3 for the year. For the following fiscal year ending December 31, 2017, Mitel reported revenues of $1.06 billion and a net loss of $49.7 million. In

November of 2018, Mitel was purchased by an investor group, Searchlight Capital Partners, consisting of more than forty partners. Searchlight took the company private and was delisted. Matthews is not listed as one of the partners nor as part of the advisory team.

Mihaylo would later say that, after the purchase, Mitel chose not to utilize Inter-Tel's VoIP technologies for which they'd invested millions.

In 1985, after the British Telecom bought controlling interest in Mitel, Matthews resigned and started Newbridge Networks. Newbridge became a global leader in the data networking industry. Sales grew to $2 billion by 2000, and the company employed more than 6,500 people. Shortly thereafter, the company was sold to Alcatel for $10.3 billion. Matthews has claimed that through his investment firms of Celtic House and Wesley Clover, he has helped spawn about 120 startups. He also has significant ownership and development in Kanata Research Park, Marshes Golf Club, a four-diamond Brookstreet Hotel, and an award-winning Celtic Manor golf resort. Matthews was knighted in 2001 and added the title "Sir" to his name.

Matthews' former partner, Michael Cowpland, upon leaving Mitel in 1984, founded Corel Corp. in 1985, supposedly an amalgamation of **CO**wpland **RE**search **L**abs. By 1997, the software, CorelDraw had achieved about seventy-five percent of the PC graphics market. In 1996, Corel bought WordPerfect from Novell, Inc. of Provo, Utah for $210 million. Toward the latter part of the '90s, Corel had reached revenues of $450 million. By the turn of the century, the company was caught in a spiral of falling sales and rising debt. Cowpland, accused of insider trading for selling shares before losses were made public, left Corel in August 2000. In August of 2003, the private equity firm, Vector Capital, bought out Corel, taking it private. On April 26, 2006, Corel completed its return to the public market. The company has since gone through a number of layoffs, restructuring, and outsourcing.

CHAPTER 20

Mercenaries: Unscrupulous Consultants and County Officials

"Get Us Girls and Get Us Money"

"Be prepared to meet me naked in a field with a bag full of money!"

That was Abe Stein's comment to Morris Jacobs, salesman for Executone, San Diego. Jacobs was telling his boss, Jim Healy, what the San Diego County Communications Director had just told him. Healy and his partner, Craig Dorsey, had purchased an Executone franchise for San Diego County a few years earlier. They had decided to hold a telecom conference displaying the capabilities of Executone products and service, and invited a number of business and public agencies to attend. It was toward the end of the conference when Stein, feigning some interest in the Executone D1000 PBX, made the above comment to Jacobs.

"What should I tell him?" Jacobs asked Healy.

"Just laugh it off. It was probably intended to be a joke," Healy replied.

However, as the day progressed, Stein became more insistent on favors from Executone San Diego, moving into Healy's office at one point and asking him to supply women for him and his boss, Larry

Gonzales. Healy replied that he didn't know any prostitutes, so Stein said, "Here, write this down." Healy added the phone number to his Rolodex and promptly forgot about it.

San Diego County had decided to go out for bids for their county-wide telephone system. A December 12, 1982, article in the *San Diego Union* written by Call Sottili, said "The plan was to install a $33 million microwave phone system [which would have also included several PBXs or switches and related telephones] to serve county buildings and departments. It was embraced by officials who saw it as an innovative answer to ever-increasing [Pac Bell] phone bills and one that would focus favorable attention on the county." Sottili added, "the idea of an independent phone systems goes back to either 1972 or 1980 depending on who is telling the story." Regardless, during the late '70s into 1981, work on the project commenced. Telink, an Interconnect company who bid on the system, was owned by a large electrical contracting company, Burnup and Sims. Burnup and Sims provided design, construction, and maintenance services to telephone companies and utilities. Telink claimed the microwave radio system and digital PBX would be able to transmit data and voice, and support more than 11,000 phones.

Stein was considered "a colorful character, given to wild claims and exaggerations." But, setting aside what one may conclude about his personality, he was apparently the one county official driving the effort. It seemed like lots of people were involved in the project—more than one might think necessary. It is unclear if Stein and Hilario (Larry) Gonzales, San Diego County Director of General Services, had much input or had established the original design specs. What is clear is the county decided to hire the consulting firm of Tele-communication Consultants, Inc. (TCI), their specialty being in radio consulting, for Phase I of a three-phase consulting contract. TCI quickly subcontracted the first phase to TDC, Telecommunication

Design Corp. The principals of TDC were Don Woodaman and Robert (Bob) St. Pierre. TDC apparently had a close relationship with Telink, an Interconnect company that sold and serviced private phone systems and who bid on the system. It would later surface that TDC would sell PCMA (Professional Communications Management Association) including one million dollars of computer equipment and services to Telink. On a "non-bid" basis, Phase II was awarded to TDC. The county claimed that of the eight bids for the county's three-phase consulting contract, only one company, TCI, responded adequately. Another asked for more time. Phase III was also awarded to TCI on a noncompetitive basis, allegedly because the county purchasing department thought they would lose too much time going out to bid and did not want to wait. Of the $459,000 county consulting contract, it is estimated that TDC received two-thirds of the fee. Although this all appears to be a rather complex arrangement involving several parties, perhaps it could be reasonably explained. Perhaps.

Once the Request for a Proposal (RFP) was ready to be mailed, county officials Stein and Gonzales set out to review capabilities and depth of the potential bidders. In short, they let it be known they expected to be wined and dined ... and more. The two were very clear about it. In the normal course of business, a prospective customer would be invited to visit telephone manufacturers, see a system demonstrated, review service and support capability, and scrutinize the distributors responding to the bid. A dinner or lunch might be included. Stein and Gonzales wanted more.

As possible bidders on the system, Healy and his partner, Dorsey, along with Executone San Diego salesman, Morris Jacobs, arranged to have Stein and his boss, Gonzales visit the Digital Telephone Systems (DTS) offices in Novato, CA. The DTS D1000 was one of the PBXs distributed by Executone. Earlier, Dorsey and partner, Bob

Knutson, had negotiated with Executone to distribute the DTS D1000 through their dealer network. After meeting DTS President, Don Green, the DTS sales manager, Bob Horvath, demonstrated the D1000 system capabilities and features. Afterward, the Executone San Diego group then drove back to the city, taking Stein and Gonzales to dinner at Perry's, a popular bar and restaurant on Union Street in the Cow Hollow district of San Francisco. All the time during the trip and over dinner, Stein let it be known that they expected some "girls." When he again requested Healy "get me some women," Healy replied, again, that he didn't know any. Before he left, Healy turned to Stein and said: "Abe, this is the hottest pick-up spot I know of in town. Good Luck." The three left Stein and Gonzales at Perry's and called it a night. They were scheduled to meet with the two of them at their hotel for breakfast the following morning. When they arrived, Stein was not there. Apparently, he had been jailed during the night. The story was that Stein either got into some confrontation with a girl at the hotel or that he was irate and demanded the hotel provide him a girl. Either way, the hotel manager concluded the best way to get rid of the problem was to call the police. When the bid process for the county phone system was turned over to a pair of consultants of questionable ethics, Healy decided that the county phone bid was going to be a stretch for his Executone distributorship and was uncomfortable the way the situation was unfolding. Executone San Diego chose to not submit a bid and thought the matter was closed. Several months later, Healy was presented with another role regarding the county bid that he would decline.

One of the competitors responding to the RFP made the comment that they were expected to pay a "kickback" to the county's consultants which, in turn, would be distributed to unnamed county officials who would make the final decision. It wasn't the county officials, but the Chamber of Commerce who first became suspicious that something was going on that wasn't quite legal. The contract for the San Diego County telephone systems including microwave connections between facilities had been awarded to Telink in June of 1982. Telink was not the lowest bidder. Anyone surprised? Recommendations for the award to Telink were made by the consultants hired by the county. An investigation commenced the following month.

Before the contract was awarded, Stein, in July 1981, resigned citing mental stress. Just what could have caused the stress? Could it have been the pressure of the large pending telecommunication contract for the county? Not getting the women he wanted? Nope. The stress was related to some of his other improprieties. Apparently, Stein had said he would personally "make a million" if the county implemented the systems. But that statement which TDC consultant tried to explain away wasn't the cause either. It seems he was the subject of a grand jury investigation for allegedly lying on expense accounts and using county workers to do chores around his home. Some of his former colleagues said that most of his comments were mere bluff and bravado. Bob St. Pierre, co-owner of TDC, provided the following explanation of Stein's comment: "When Stein said, 'I'm going to make a million dollars,' what he meant was, 'as soon as this system goes [in]... state governments, county governments, huge corporations are going to be looking for people who have already done it. And I can go in there as a consultant or sell my services to a manufacturer as a special liaison person. And I can command big, big bucks.'"

Maybe some officials bought that explanation, but most did not.

The Greater San Diego Chamber of Commerce, Pacific Telephone, county officials, and several of the firms that bid on the contract started hurling accusations at one another, claiming kickbacks, extortion, and corporate sabotage. Investigations into the selection process ensued and involved the county grand jury, the district attorney's office, the FBI, and the U.S. attorney's office. In 1982, the district attorney's investigators said $12,000 in $100 bills were found in a safe-deposit box belonging to Stein. The probe was first launched before a county grand jury, and in January 1983, it shifted to a federal grand jury as alleged criminal violations were more closely aligned with federal statutes. Even though a small part of the installation had begun by Telink, the contract was put on hold while the investigation continued.

Thomas H. Bell, the salesman on the west coast for Universal Communication Systems (UCS) who had bid on the county contract slid comfortably into his chair possibly thinking to himself, *I've got this.* Bell was described by his co-workers as on the short side, high energy, somewhat heavy set or stocky, blond hair, a bit flashy, well dressed and very confident—"a ladies' man." At the time, he had a house near the point of Tiburon—one of the more expensive communities in the Bay Area across from San Francisco. The day of the trial, in his good suit, his confidence was on display. He showed no hint of being nervous before the grand jury. His answers to the questions asked by the prosecuting attorney were succinct.

Did he receive any bribes in the process of submitting a bid for the county contract? Bell answered crisply, "No."

Did he receive money in the form of fees or inducements? Another "No."

"Didn't you receive two cash payments of $5,000 each from the consulting firm, TDC?" asked the prosecutor.

"That was a loan." Bell answered.

The prosecuting attorney hesitated. "I noticed your watch. Is that a Rolex watch you're wearing?"

Bell proudly answered, "Yes," and when asked how and where he got it, he lied again. It would have been most unusual for any consultant to loan any money to a salesman with whom they had limited business dealings—especially an amount as large as $10,000.

The San Diego Board of Supervisors canceled the $24 million contract in January 1983, after the grand jury began the investigation which was later turned over to federal authorities. No explanation was provided as to how the county contract went from $33 million as first reported in local papers to $24 million.

As the investigations continued, an investigator for the District Attorney interviewed Healy in his office after interviewing other Executone San Diego employees. Healy confirmed the trip to Novato and San Francisco with Gonzales and Stein, and that Stein had pressed Healy into procuring some women for the two county officials. Healy informed the investigator that he had not acceded Stein's request.

"Do you know Miss 'X'?" asked the agent.

Healy replied he did not.

"That's your Rolodex?" pointing to the business card filing unit on Healy's desk. "Let's look." The agent proceeded to pull out the scrap of paper that Stein had provided to Healy with the name and phone number of the lady in question in Healy's handwriting.

"Whoops," Healy exclaimed. He had forgotten about it. But one of his employees must have remembered, as the investigator knew where to look. The investigator appeared satisfied with Healy's responses and asked if he would be willing to wear a "wire" and meet again with Stein. Healy made a quick phone call to his attorney who advised him to refuse the request, which he did.

On November 1, 1984, the 107-page indictment was announced following the arrests of thirteen people mentioned in them. Charged were Hilario Gonzales, Abe Stein, Don Woodaman, President of TDC, Bob St. Pierre, TDC VP, Telink, Inc., Bobby Hendricks,

Telink President, Burnup & Sims, Inc., Telink's parent, James Linder, Telink Marketing Director, Robert Palmer, manager of the Telink San Diego office, John Bostwick, Telink Controller, Henry Richter, President of Telecomm Consultants, Inc. (TCI), Bernard Campbell, TCI VP, Robert Schreiber, TCI VP and lead consultant on other projects in Fresno and San Diego, Michael Sage, VP, Secretary and Treasurer of TDC, and David Stein, Abe Stein's brother. Two companies were also charged. The indictment stated that Linder and Stein "paid prostitutes to perform sexual acts" for Gonzales at Stein's home and elsewhere. It also accused Linder and Stein of recruiting a nineteen-year-old Telink employee as a prostitute to perform sex acts "in order to influence the actions and conduct of individuals responsible for the award of the San Diego County contract." The indictments also charged that some defendants encouraged others "in the distribution and use of narcotics," so that those people—all connected with the awarding of telecommunications contracts— could be blackmailed. The indictment did not say who was involved in these alleged incidents, citing only "unnamed officials." The indictment charged one count of racketeering, one of conspiracy to engage in racketeering, thirty-eight counts of mail fraud, two of wire fraud, three of obstruction of justice and four racketeering-related counts. District Attorney, Edwin Miller termed the case "the most massive fraud and public corruption scheme ever perpetrated against the county of San Diego." Prosecutors also discovered that the conspiracy and illegal activities extended to other contracts— including one for the county of Fresno, plus numerous other contracts with private business. Tom Bell was not named in the indictment. He was already in jail for lying to the grand jury.

An unidentified woman whom Tom Bell had kissed when first entering the courtroom began screaming. With tears streaming down her face, the woman shouted to U.S. District Judge Earl Gilliam:

"No! You can't do this to him. He didn't know anything about what was going on. People have lied on him ... Oh, God, no, it just isn't fair." As a bailiff led Bell into custody, she cried, "He didn't do anything. He's such a good person."

Outside the courtroom, she refused to identify herself. Peter Campbell, UCS Western District Manager, remembers. He had recently been transferred to UCS's Southern California office in Fullerton to "clean it up." Campbell had driven Bell's girlfriend, Athena, to the courthouse to hear the sentencing. Bell was given a sentence of three years in prison. At the time of his sentencing, federal prosecutor Charles Hayes insisted that Bell was still withholding information from the continuing grand jury investigation. After about ninety days in the Lompoc penitentiary, Bell decided to "come clean" and volunteered that he had lied under oath. Bell admitted that Woodaman and St. Pierre, owners of Telecommunication Design Corp. (TDC), had asked his firm for $500,000 for a favorable recommendation on the contract. Bell said that unspecified portions of that amount were to be paid to Gonzales, former county General Services Director, and Stein, former Chief of County Communications. Gonzales was fired in late 1982 for alleged misconduct in connection with bids on the phone systems. The gold Rolex was allegedly provided to Bell for inflating the price of the UCS bid so as to make the award to Telink less conspicuous.

According to the indictment, "The defendant, Telink, maintained photographs of defendant St. Pierre in the company of a nude female prostitute during the period that St. Pierre served as consultant primarily responsible for recommending the award of San Diego's $24 million contract to Telink." The indictment also alleged that Telecommunication Design Corp. (TDC) engaged in a pattern of bribery and fraud to obtain contracts for the installation of telephone systems at sixteen different locations throughout the

West, including, locally, Coronado Hospital, Burroughs Corp.,
Home Federal Savings, and the city of San Diego. District Attorney
Miller said they would seek $10 million in forfeitures from Burnup
& Sims (through Telink) for allegedly engaging in illegal acts to
obtain contracts from the Fresno county, Glendale Federal Bank,
Burroughs Corp., and U.S. Borax.

The indictment described an elaborate money laundering system.
Bank accounts were set up under phony names. In one example, a
Telink check was allegedly laundered through TDC, allowing
Bostwick, Telink comptroller, to carry $60,000 in $100 bills in a
brown paper bag for a drop-off at the Bob's Big Boy restaurant in
Oceanside, California. Stein, who picked up the cash, mailed it to an
unidentified address. Considering Morris Jacobs' earlier experience,
one wonders if Bostwick had to be naked. The $60,000 delivered to
Abe Stein in August of 1982, the indictment alleges, was in
connection with the rigging of competition for a $4 million
telecommunication contract with Fresno County.

Telink didn't take the investigation passively when the county
cancelled the contract in January of 1983 after the release of court
affidavits. They filed a $57 million claim for breach of contract
against the county, but the threat of legal action ended when the San
Diego county paid them a settlement of nearly $2.5 million
including a $700,000 profit for Telink and $500,000 in penalties
Telink faced because of the contract cancellation. It is doubtful the
county ever requested verification of the cancellation fees.
Traditionally, if the manufacturer hasn't shipped the system, the
order could often be cancelled without penalty. District Attorney
Miller said that settlement had been a mistake, saying, "had they
sought our advice, we would have vehemently opposed a settlement."
Clifford Graves, the county's chief administrative officer rationalized
that the "settlement was prudent at the time." Graves claimed that

Telink had installed telephone systems in the county's South Bay and East County offices which were in use and "working quite well."

Not all telephone consultants were unethical. Many started their carrier with the Bell System and tended to primarily recommend Western Electric systems. That attitude would change as other manufacturers came to market with valuable cost-saving features. However, in the '70s and '80s, the San Diego County Telephone bid was only one of several consultant-involved bid rigging schemes across the country. Additional instances are also mentioned in following chapters.

● ● ●

BOOK THREE

Major Players Stumble, Others Succeed

All who entered the business during the '70s were optimistic about their future.
By the middle of the decade, several new entrants would appear.
The late 1970s and 1980s would include some unexpected challenges.

CHAPTER 21

TIE Communications, Inc.: The Rose Fades

D. H. Blair, who first invested in the MEC startup now incorporated as TIE, turned over their investment to Allen & Co., who sold their investment in TIE, eventually buying it back. D.H. Blair was the leading underwriter who took TIE public. The initial offer went out at $5.50 a share and was over-subscribed.

TIE's position in the market began to be diluted with new entrants—some offering expanded features. Once only a fraction of TIE's size, Inter-Tel expanded their distributor base and added features normally found on the more sophisticated PBXs. Other manufacturers, wanting a piece of the action, attracted dealers who were looking for any product other than TIE, as every competitor in town was a TIE distributor and dealer margins were waning. To make matters worse, many of the small Interconnect dealers during the last part of the '70s and into the first half of the '80s were faced with the highest interest rates in U.S. history. The prime lending rate in 1980 climbed to twenty-one and a half percent, the highest prime rate ever recorded, placing pressure on both dealers and suppliers. TIE's receivables grew. Until closing their offices, Litton BTS had become probably their largest distributor. According to Sherman,

"When Litton pulled out of the market in 1974, TIE had two huge warehouses filled with TIE systems." Sherman and Engles were able to off-load most of the Litton inventory to Executone, Inc.

Bob LaBlanc, an analyst at Solomon Bros., called Kelly, saying to him, "You've got a problem." Referencing Litton's suit with AT&T, he added, "You don't want to get involved with a lawsuit with AT&T." How TIE might be damaged through the Litton suit or how an antitrust action initiated by TIE against AT&T might negatively affect the price of TIE's stock can only be speculated. LaBlanc added that he thought litigation might be avoided through a settlement, and set up a lunch meeting with Charlie Brown, the AT&T chairman who succeeded John deButts. As Kelly recalls: "Lunch was at the top of AT&T's head-quarters at 195 Broadway. It was a wonderful room and a nice lunch. During the lunch LaBlanc said to Brown, in relating to the Litton suit: 'I'm going to have to sue and we don't want to.' ... After the lunch I told LaBlanc I didn't know what that was about. Bob said to me, 'I think we can get you a settlement.' AT&T offered (1) paying $500,000 in legal fees, and (2) a settlement to TIE of $1 million."

TIE took the offer. It is questionable whether the TIE share-holders would have benefitted more by TIE's filing an antitrust suit against AT&T or joining Litton's suit. However, "take what you can when you can" wasn't a bad strategy, as TIE may not have had that opportunity at the time.

The financial situation at the time was stressful for those competing with AT&T. Other AT&T activities would place additional pressure for all Interconnect competitors. AT&T had been lobbying effectively with congress to get Asian import duties increased on telephone equipment, effectively raising the cost of most all Asian-made tele-phone systems. Then, during the last part of the '80s, the industry experienced a significant market downturn, dropping some 500 points in a day.

American Business Phones of Irvine (ABI), led by founder Frank Feitz, had taken their growing company public. In 1987, in need of more capital, ABI agreed to be purchased by TIE. The acquisition was announced in the *Los Angeles Times* on December 12, 1987. ABI had reported $27.1 million in sales with a profit of a little over $1 million. In contrast, TIE reported a loss of $59 million, including an inventory write-down of $34 million on sales of $298 million for their fiscal year ending December 31, 1986. For the first nine months of 1987, TIE had losses of $4.3 million on sales of $191 million.

TIE reported revenues of $100.5 million ending year 1990, and incurring $8.7 million in losses. As Steve Mihaylo, CEO of Inter-Tel, observed, "TIE had about one full year of inventory in the late '80s when we were turning our inventory every month." He concluded that may have been a contributing factor leading to TIE's downfall and added, "TIE also made an effort to go after [selling to] the Baby Bells after divestiture, but the market size didn't change—there were the same number of potential key system customers before AT&T's divestiture."

TIE filed for Chapter 11 in the U.S. Bankruptcy Court on April 8, 1991.

Engles said that he thought one of the Pritzker brothers had become involved with TIE before the Chapter 11 filing. Regardless, a plan of reorganization submitted to the U.S. Bankruptcy Court by Marmon Holding, Inc. and the Pritzker Family Philanthropic Fund was approved on July 1, 1991. The plan was to move the new company to Chicago, home of the Pritzker family. Engles said he wasn't about to move and resigned.

In 1991, the reorganized company announced the discontinuance of its distribution business. They would no longer be involved with research and product development. The emerging new TIE changed their business model to one more resembling an advanced Interconnect distributor offering PBXs and ancillary voice and data products: voice response and processing, video conferencing systems, long distance products, computer telephone integration hardware and software, as well as pre-owned (used) phone systems along with lease and rental programs. Their 10K filing of 1993 sounded like the new TIE would provide everything—they just wouldn't make it or distribute it.

The company entered into a nonexclusive license agreement with NTK America, owned by Nitsuko who had provided product for TIE for twenty years. The reorganized TIE also acquired PacTel Meridian Systems' (PTMS) base of 14,000 customers for $7 million, less $3.4 million for liability reserves for unexpired warranties and service contracts. In 1993, the company reported revenue of $102 million and a profit of a meager $1,365. The profit included royalties (expiring in 1996) of $1.378 million and $1.74 million resulting from the sale of the previous Connecticut building. The Marmon Group moved TIE to Overland Park, Kansas in 1993. During the year, TIE also sued NYNEX for a breach of contract over the sale of NYNEX's customer base. NYNEX instead decided to sell its customer base to WilTel, a TIE competitor. The court decided in favor of NYNEX. The following year, the company reported revenue of $125 million, but losses of $3 million. That year, the company filed a Notice of Termination with the SEC and was no longer listed. Bloomberg Business on December 10, 1996, announced a three-year agreement with TIE and Phoenix Network, Inc., a reseller of long-distance service. The *Kansas City Business Journal* on February 28, 1998, stated that in 1995, TIE's remaining assets had been sold to SP Investment, a Seattle-based investment company. TIE's meteoric rise had fizzled.

CHAPTER 22

TeleResources, Inc.: Former IBM Execs Create a Ground-Breaking Product

The Armonk and White Plains area of New York is home to IBM. A few former IBM managers in the late '60s created EDP Resources, a leasing company. Aware of the FCC *Carterfone* decision, management wanted to expand their leasing business and decided that leasing telephone systems would provide considerable growth for the company. They concluded the best way to get into the market would be to start an Interconnect company selling and installing private telephone systems. We don't really know that was the primary motivation, but W. Brian Satterlee, Jr., President of the leasing company, followed the suggestion of fellow IBM employee, Charles Pedler, and created a new subsidiary, TeleResources (T/R) in 1969. The plan from the beginning was to design and build their own PBX product. In order to gain the required experience about the market and what the customers would want, they decided to operate the first few years as a typical Interconnect distributor, selling other PBX systems while designing their own. Revenue from the Interconnect operation would also provide funds for the new company. Satterlee brought Pedler with him, who would become the

President and chief architect. Bill Jacobson was one of the first to be hired by management.

Jacobson, upon graduating from college, had been hired by Southern New England Telephone Company and had worked in their marketing department for seven or eight years. Feeling a little confined working for a Bell operating company, Jacobson left to join another startup company that was in the timeshare business. "It was a shock! I had been working in a regulated industry where ethics and honesty were essential. The timeshare company was a bunch of crooks. Not everyone was honest—the opposite of the telephone company. It was a nine-month learning experience for me," Jacobson recalls. "I was going to be more cautious in my next career move, placing effort on knowing the background and quality of the people I would work with. 'What are the people like at TeleResources?'" Jacobson had the industry background that Satterlee and Pedler needed. Pedler, ten years senior to Jacobson, was a graduate of Hamilton College in Clinton, New York, the same small private liberal arts college Jacobson attended.

"Pedler was smart and technically savvy—but not one I would call a driving CEO," Jacobson said. "Before developing the TR32 [PBX], we began evaluating both Stromberg-Carlson and OKI PBXs. It helped that Stromberg's offices were in nearby Rochester. We decided the OKI would be more cost and feature effective. Selling, installing, and servicing other PBXs gave us confidence in designing and building our own product. Charlie did the technical design and created overall specs, while I defined the attendant and station functionality. I wanted to name it 'New Dimension,' but was voted down." In the late '70s, AT&T's Western Electric would later develop an electronic PBX called "Dimension." "Instead, we called it the TR32, as it could support thirty-two simultaneous conversations. Development of the final product was outsourced to a firm in North Carolina. Pedler, a private pilot, and I would fly in his plane to the North Carolina facility, which was sometimes scary. We were the first company to build and market a completely electronic PBX. Initially, the development was funded internally."

Although not a fully digital system, by comparison, AT&T's Western Electric's most modern PBX in the same target market was the 3a (or 4a console) 756. It only supported ten to twelve links for simultaneous conversations and offered a smaller number of trunks and stations. However, the transfer of calls and intercom calls also required links. With any heavy traffic, the 756 would become strained. Smaller in size to the Western Electric products and with a more attractive console (attendant position), the TR32 also employed a single line telephone with advanced functionality. The telephone housing was similar in appearance to Western Electric's Touch-Tone™ single line telephone with an extra feature button/key. By depressing the key, the user could place a call on hold, make another call, transfer an incoming or outgoing call, or create a conference call—features that all Western Electric telephone systems at the time lacked unless the attendant would perform the function.

When Bell System customers demanded more capability, the common practice for the Bell marketing department was to add expensive key telephone systems behind the PBX. Even so, that solution still didn't provide the features of the TR32. The TR32 telephone was absent of a lot of hardware employed in Western Electric's single line telephones. Gone was the ringer (bell), coil, and heavy base. Instead, the TR32 telephone only contained a small circuit board to generate the DTMF tones—it was very lightweight. There was one small problem: If the user picked up the handset too quickly, the tension in the coil cord between the handset and telephone instrument would cause the telephone to creep across the desk toward the user. Infrequently, a customer might complain about the problem. A quick fix was placing a lead weight in the phone to keep it in place.

A small electronic speaker, similar to those used in phones used today, was employed to generate the ring tone. Unfortunately, when a power outage occurred, all the telephones in the system would emit a "cry" alerting everyone to the fact that the system was down. Regardless of these two minor points, the TR32 was an instant success.

Final development and beta testing of the TR32 was contracted to a company in North Carolina. A little over two years in development, the TR32 was introduced in 1972. Jerry Powers handled the marketing, and the company engaged Stan Blau to appoint distributors. Dick McGuire headed up the sales effort. Ray Charbonneau, another NYTel alum, came on board shortly thereafter. The latter two opened a sales office in Long Island where both lived. TeleResources decided to sell the TR32 direct in the New York metro area as management wanted direct feedback on market acceptance, reliability, service requirements, etc. Another of T/R's early salesmen was Pete Cunningham who worked for a year as a marketing consultant at NYTel. A fellow employee told him that her nephew had just gone to work for an Interconnect company and suggested he contact them. After being hired by T/R, Cunningham recalled one of his first customers: "He asked how much the system cost per pound, as he only purchased things that cost less than $1.00 per pound. I didn't know, but returned to the office to weigh everything, including cable, terminals, phones, and switch. When I came back and told the customer it was less than $1.00 per pound, I got the contract. During the installation of another TR32, a NYTel division manager saw that a private telephone system was replacing one from AT&T and told me 'You cannot do that!' He began reading the 'riot act' to me. Fortunately, I had some articles with me on the Carterfone case that I showed him. He was not happy. As TR started manufacturing their own product, I was moved from sales and given responsibility for parts and labor at the manufacturing plant."

Alan Kessman was hired in 1973 as the company's CFO. He recalls, "At the time I was hired, TeleResources was doing about $8 million in sales and losing money. When I resigned in late '77, the company had grown to $75 million in sales and was profitable. Pedler was President when I joined as CFO. As EDP Resources was going under, TeleResources was spun off and Satterlee, who was CEO of EDP Resources which held all of the TeleResources stock, took the position of CEO of TeleResources, and Pedler became VP of Product Development." Satterlee didn't want the daily responsibilities of running TeleResources. "He let me basically run the company. We had a gentleman's agreement that I would receive compensation and stock at eighty percent of whatever he received as CEO. As evidenced in a September 23, 1975, memo from Satterlee, he announced, 'the election of Mr. Alan Kessman as Executive Vice President and Chief Operating Officer.' When we turned profitable and [the company] was humming along, he came to me and told me that the Board had met without me—even though I was a member of the Board—and decided to raise his salary and bonus but not mine. He also said that our agreement whereby I would become President and CEO was no longer valid. I made it clear that the Board could not hold a meeting without even notifying me. At that time, the other Board members were the former Board members from EDP Resources and friends of Satterlee so my objections fell on deaf ears. I then gave Satterlee one year's notice to replace me, and I told him I could not work with people I could not trust."

When asked why Satterlee chose to disregard his commitment, Kessman responded that, "Satterlee told me that their verbal agreement was no longer valid and that he was taking control back because the company was doing well, and he wanted the credit for it." Broken agreements are often remembered more clearly than those that are kept.

Fortunately, TSI was one of the first authorized distributors. The TeleResources distributorship provided a protective territory, enabling us to maintain a good profit margin. The TR32 would become TSI's most profitable product.

In the same September 23, 1975, memo, announcements of William (Bill) Jacobson's promotion to VP of Planning and Gerald (Jerry) Powers's promotion to VP Distributor Marking were made. At the time, T/R had built the distributor network to over forty. The TR32 went through three successful versions, receiving favorable publicity when *BusinessWeek*, on September 6, 1975, published an article, "Tallying phone calls a cheaper way," about TeleResources. The article introduced Call Detail Recording (CDR) of outgoing calls that generated reports for cost accounting purposes, and allocating phone costs to departments. T/R claimed that management's use of a CDR system could eliminate up to ninety percent of personal phone calls. T/R was not the first to market such a product, but other CDR systems made by ESE, Ltd. of Toronto, Danray, Inc. of Dallas, and Communications Equipment Corp. of Hartford were much more expensive and aimed at a larger market: companies with toll charges of $6,000 to $10,000 or more per month. The TR32 CDR only added about twenty percent to the cost of their standard PBX. A little over a year later, ROLM would announce their CDR version. Regardless of the favorable positioning of the TR32 and its cost effectiveness, the company was running short of cash. Satterlee was searching for new funding, but time was running out.

Shortly after Kessman left T/R, his friend, Stan Blau, made a call on his behalf to Ken Oshman, President of ROLM, asking if they had a position for him. The two met, and Kessman was hired during the summer of 1977. While at ROLM, Kessman received a call from Bill Hambrecht of Hambrecht & Quist, investment bankers.

Hambrecht told Kessman their firm was considering making a $5 million investment in TeleResources and wanted Kessman's opinion of the company. Kessman agreed to visit T/R and would report back. "I visited T/R and did a complete due diligence [and] told Bill Hambrecht that Satterlee had ruined it again, and I thought it was too far gone to be saved." Hambrecht added at the end of the conversation that if Kessman ever decided to do something on his own and needed funding, to call him—an invitation that Kessman would not forget.

Unable to secure the required funding, TeleResources filed for Chapter 11 in late 1979. But a plan for reorganization was never submitted to the Bankruptcy Court. Several reasons may have contributed to their demise. A major cause may have been that the system was priced too low, and that management failed to increase the price to its distributors as sales increased. The company needed more money and time to continue its growth, and had neither. An expansion to Ballston Lake in upstate New York near where Pedler lived added to the company's overhead. Also, the TR32 was not a digitally-based system. The industry was changing to microprocessed controlled digital technology that was more capital intensive, and the architecture of the TR32 was not capable of being digitally upgraded. Competition also played a role in putting stress on the company. Mitel, Siemens, and OKI PBXs had become favored products with many distributors, and both could support conventional key equipment behind their PBXs which the TR32 could not. Both Mitel and OKI were targeted at the same market as the TR32. Last, there is the question of management. While Kessman was in charge, the company was generating capital. Upon Satterlee's return and Kessman's departure, profitability declined and the company ran out of cash.

CHAPTER 23

Executone Information Systems, Inc.—"Let's Sell It"... and Sell It Again...and Again

E xecutone had decided to increase its market and add more revenue by offering telephone systems to its dealers to sell to end users. This wasn't quite as easy as a supermarket putting another product on its store shelves. Not only were the Executone dealers not trained in telephone systems, the Executone management and its staff were not in a position to provide sales support for telephone products. Worse, when it came to PBXs, the company did not have the resources to provide technical help to their dealers. One former Executone dealer reported that, in order to make a statement about his unhappiness with the lack of support, an Executone salesman, Tommy Childers, parachuted unannounced into Executone headquarters in Long Island. We don't know if his message was heard by senior executives, but it's possible his career with them was either dramatic or short lived. Small key systems acquired by its dealers through Executone were more expensive than if the dealers bought directly through the same supplier with only a slight cosmetic change— which some eventually did. A few Executone dealers actually formed separate companies that purchased non-Executone provided phone systems to avoid violating terms of the distributor agreements. The

D-1000 PBX from Digital Telephone Systems provided interim relief as it did not require the level of technical support of previously supplied Executone PBXs.

By 1966, Continental Telephone Corp. (Contel) had become the third largest independent telephone company by acquiring 500 independent telephone companies. Seeking to secure a position in the expanding deregulated market in 1979, Contel acquired Executone whose annual revenue at the time was reported to be just under $200 million. The company headquarters was moved from Long Island, New York, to Atlanta in 1985. A web-based service that publishes on its site just about anything any company wants to say about itself, its history, and products, cites the move to Atlanta as one of the main reasons contributing to Executone's "poor performance" for the year. The same publication also speaks to the loss of personnel, including many of its senior employees and a rapid turnover in top level management with the replacement of six presidents in nine years.

Executone reported a loss in 1987 of $15 million. Contel had become disillusioned with Executone and decided to sell the subsidiary. The same year, two of Executone's primary suppliers—Vodavi Technology founded by Steven Sherman, and ISOETEC Communications, a relatively recent company established by Alan Kessman—agreed in December 1987 to jointly acquire Executone for $60 million. The following July of 1988, shareholders approved the acquisition, and the surviving company was named Executone Information Systems (EIS). Kessman became President and CEO. The Vodavi division continued to operate in Arizona. Executone headquarters, when bought by Contel, had been moved from Long Island to Atlanta. With the new three-way merger, Kessman moved the company back to Milford, Connecticut. Only a small percentage of employees were offered positions with the new company.

Between the date of the informal agreement and its finalization months later with the three entities, some of the ISOETEC and Vodavi senior management began to move into their new positions in EIS. Kent Burgess, as the Vice President of Production, had responsibility for purchasing, inventory, and technical support, the same position he'd held with Vodavi. He remembered visiting Executone's Long Island facility and commented, "All I saw was older people in their 60s. They once made their own product in nurse call and hospital systems, but I didn't see any activity at innovation. I went to check their warehouse. An older fellow was asleep behind his desk, and I had some difficulty waking him. When he woke, he looked at me saying, 'Who the hell are you, and what do you want?' It was a completely different attitude than what I experienced at Vodavi."

With the purchase of Executone came some serious challenges. In any major city or metropolitan area there would likely be (1) an Executone distributor or direct office, (2) an ISOETEC distributor, or direct office such as recent ISOETEC acquisition Jarvis Corp., (3) one or more distributors or electrical contractors selling Vodavi systems through Graybar or North Supply. This was at a time when the market was beginning to become crowded with product. It was confusing, with salespeople stepping on each other's toes. Nick Visser, former President of Jarvis Corp., became the VP and General Manager of Executone Info. Systems, Jarvis Division, reporting to Kessman. Visser was taxed with consolidating some fifty-plus offices down to a more manageable thirty.

Visser recalls, "I found Kessman pragmatic, very organized, and disciplined. However, the merger was painful. Initially, I had the profit/loss responsibility for fifty offices. We needed one set of policies and procedures, and we needed to consolidate. EIS wanted to have one office in each city with a branch or general manager, and the choice I had to make was difficult. On one occasion, I selected an

Executone manager over a competent former Jarvis branch manager whom I knew well. She was very upset with me. Some monthly meetings lasted an entire week. By 1994, I was getting really tired and thought of leaving."

The Executone management entry in the online business database mentioned earlier waxes on about "business philosophy... vertical integration strategy of complete control over product design, manufacturing, market, and services while... diversifying its product line and expanding its end-user base for post-sales activities... reduced competition through consolidation... elimination of duplicate resources... etc." The comments sound like the company was positioned to hit the streets running and make lots of money. However, in its first full year of operation in 1989, the company again failed to make a profit. They blamed reorganization issues and the U.S. Commerce Department import duties "as high as 158% for several Far East Asian multiline business communication equipment manufacturers, including Korea's Goldstar Telecommunications and Oriental Precision Co., two of Executone's major subcontractors." Steve Mihaylo and I both recall import duties levied at the time were in the midteens. As such, the 158% may be a typo or applied to specific products other than telephone systems. During the period of 1991 to 1994, the company demonstrated small but consistent revenue growth from $244 million to $292 million in 1994. However, during the same period, the company only managed to make a small profit each year, ranging from .4% to a high of 2.5% on revenue, or $7.5 million in 1994.

Additional services of long-distance dialing through their Infostar LD+ and new features of automatic call distribution (ACD) and predictive dialing (automatic out-bound dialing) were also introduced in 1995. Some of these features had been introduced by other telephone manufacturers years earlier. Phone systems with ACD had been installed by a few manufacturers in the late '70s and early '80s. To top things off in painting a rosy picture for the future, the company touted its Sales Navigation and Computer-Aided Selling (CAS) programs that streamlined and enhanced the selling effort required in the market. While the programs may have contributed to the company's profitability during the '91-'94 period, the company reported an operating loss of almost $37 million in 1995. Part of the loss was attributed to the purchase of certain Vodavi assets by Sherman and Fitchet the previous year. In December, the company announced the acquisition of eLottery and its subsidiary, UniStar, a national Indian lottery business.

On April 11, 1996, *The New York Times* reported Executone sold its telephone sales, service, and long distance business to Bain Capital Inc. and Triumph Capital Group for $65 million. Bain Capital, founded and run by Mitt Romney, made investments and purchased hundreds of companies. An online search of Bain Capital's investments and acquisitions did not include Executone Information Systems. Several EIS executives stated that Bain's intent was to merge Executone with Staples, another of their acquisitions. Personally, I didn't see a fit, nor did several others in the industry. These were completely different markets. At the time of the purchase, news articles stated that the part of the business remaining after the sale of their telephone sector would continue to pursue the lottery business, headed by Kessman.

"We were ahead of our time," recalls Kessman. "The Indian reservations had rights to establish gambling casinos on their property, and we had been working with the tribe in Coeur d'Alene, Idaho, with the intent to establish a national lottery system which would be available over the internet."

Kessman continued as president in '96 and '97, but resigned in June of 1998, taking $1.246 million under his employment agreement for early termination. The company's 10-K financial statements for the year also showed that he owed the company $2.4 million relating to an Executive Stock Incentive Plan.

In October of 1999, Inter-Tel reached an agreement with Bain to purchase Executone's distributor network and assume some liabilities and distributor agreements for $44.3 million. Regarding eLot, the Indian lottery business Kessman intended to pursue, it had little revenue for 2000 and the first nine months of 2001, reporting an operating loss of $23 million at the end of the third quarter 2001. The company cited legal difficulties with AT&T which had challenged their legal ability to operate a national lottery. The following month, on October 15, 2001, the company filed their 10-Q with the SEC showing an accumulated deficit of $106,685,000 and also filed a voluntary Chapter 11 petition with the U.S. Bankruptcy Court, Southern District of New York. ELot filed a Notice of Termination of Registration with the SEC on December 31, 2002. Although the company as Debtors-In-Possession indicated an intent to submit a Plan of Reorganization to the Bankruptcy Court in early 2002, it appears none was filed. End of eLot. After the Executone Information Systems dealer network was sold to Inter-Tel, Inter-Tel was sold to Mitel. Several former Executone distributors still continue to operate today under the Executone name selling Mitel and other products. Executone has been sold four times and merged into different companies. A once nationally recognized company with 2,400 employees servicing a network of 200 direct and independent distributor sales and service offices no longer exist.

CHAPTER 24

Litton Business Telephone Systems: A Very Short Time in Business Pays Off

AT&T's defense in Litton's antitrust suit held that Litton went out of business as a result of mismanagement, incompetency, and dishonesty—not as a result of AT&T's antitrust violations. Although AT&T made their case about Litton BTS's shoddy business practices and "bid rigging" indictments, they did not pursue individual testimony from those conducting interviews of the alleged fraudulent acts—acts that included offers of bribes, "kick-backs," finder-fees, girls, and "anything to make the sale." Perhaps AT&T feared additional exposure. While the court had agreed to allow AT&T to present its testimony providing that Litton would also be allowed to enter AT&T's acts of bribery into evidence, the court also imposed sanctions on Litton. Their in-house counsel had been found grossly negligent for stating during Norman Roberts' deposition that "there were no other relevant documents." It was later discovered that notes of Roberts' interviews with five Litton BTS employees had not been turned over. Litton maintained that the investigation directed by Litton BTS new President, Robert Bruder, was directed only at the San Mateo matter, when in fact San Mateo was only one of seven instances of suspected misconduct that he'd wanted investigated. As a result, the court denied Litton recovery of all costs and attorneys' fees to which it would otherwise have been entitled.

The evidence and testimony AT&T entered wasn't enough to sway the jury decision in the first civil case. On June 30, 1981, *The New York Times* declared: "Litton wins $92.3 million suit against AT&T." The awards were based on special jury findings that AT&T used its telephone monopoly illegally to monopolize the telephone terminal equipment market, thereby excluding Litton as a competitor and imposing costs on Litton as a customer of the AT&T system. As this was an antitrust case, if the decision was upheld, that could result in triple damages. AT&T appealed the verdict in the U.S. Court of Appeals for the Second Circuit. The case was argued on June 14, 1982, and decided February 3, 1983. The jury found that AT&T filed interface tariffs in the first instance in bad faith. It also concluded three practices of AT&T were "anticompetitive and predatory": (1) opposition to certification (in lieu of PCAs), (2) delays in providing PCAs, and (3) conduct in connection with the sale of their inside wiring. AT&T professed a willingness to sell wiring but negotiated in bad faith by quoting unreasonably high prices. As one Central Bell manager noted, the "practice of destroying inside wire was unreasonable and could be interpreted as vindictive." That was the conclusion the jury reached, and the court saw no reason to overturn it. **Litton was awarded $276.7 million in damages.**

CHAPTER 25

ROLM: The Battle Won; Stockholders and Employees Reap the Rewards

R OLM quickly became a leader of innovation in the telecom industry. One of their particular strengths was the way they brought their distributors together. Senior management at ROLM made an effort to keep their distributors motivated. Those of us who had distributed other products would occasionally attend one-day meetings hosted by companies like TIE and OKI, usually at their offices. You didn't always feel like attending, but you felt obligated to make the trip. Right from the start, ROLM raised the bar. The meetings were not only well organized, most of the time they were held at exotic places—Cabo San Lucas, Puerto Rico, Saint Lucia, and Hawaii to name a few. And not just for one day. Everything was first class. Meals were fantastic, and the room cost and incidentals were covered. One unsuspecting outcome of the meetings was the relationship that developed between several of the ROLM distributors. Many of us would make arrangements to stay before or after the meeting and enjoy each other's company. Some of us would meet together for short vacations during the year at ski areas. These relationships were the seedbed for our informal ROLM Distributor Group. We didn't get the feeling that ROLM liked the idea of its distributors meeting separately, but we did share common goals and concerns.

ROLM's rocketlike ascension in the marketplace did not go unnoticed. Very large players like AT&T, Northern Telecom, and Siemens, plus some newly funded tech companies, had either hinted at or announced plans to develop competing computer driven PBXs. Even IBM had talked about the prospect of entering the business. Later entrants, while still having to develop their own software, had the advantage of starting with more advanced hardware components. Even in the mid-'70s, ROLM management was looking over their shoulder. From the very beginning of the CBX project, they were converting voice to data, digital "0"s and "1"s, and back to voice. The CBX made the conversion in the switch at the time which was already switching data. Adding data capability would offer in-house data managers longer transmission to the mainframe or host while reducing expensive multi-conductor cabling and modems needed at the time. ROLM made the announcement to provide both voice and data for the "Office of the Future" at the 1979 ICA trade convention. Also at ICA, ROLM demonstrated their first stab at an Electronic Messaging System (REMS). With the exception of a few companies, like IBM, the only way companies took phone messages was on those pink "While-You-Were-Out" pads. The system allowed digital voice messages to be recorded, transferred to others, and accessed remotely— a big timesaver for both large and small companies.

A test bed was installed at General Electric's office. The message system used ROLM's CBX back-up computer. But when that computer was needed, it dumped all its messages. GE employees had been hooked on the system and were furious. It was removed and would not be reintroduced until 1982 as PhoneMail, which became a huge success.

ROLM had realized that they would need to upgrade the CBX.

But technology had improved, so the challenge became how to upgrade its existing base. A promise ROLM had made to its customers was that their investment would always be upgradable. While they worked on a separate development effort, *BusinessWeek* published an unfavorable article. It pointed out that since ROLM's announcement in 1979 to offer products of the "Office of the Future," the company had failed to produce any such products and that its seven-year-old CBX was now facing competition. ROLM's stock had plummeted and margins were decreasing. The company needed something to "leap-frog" the competition and regain its reputation in the industry. The answer was the CBX II—one system from forty to 10,000 lines. ROLM's CBX II announcement in November of 1983 stated that, "the 13,000 systems we've already installed can be upgraded to CBX II capabilities very easily, very painlessly, and very, very cost-effectively." The press picked up on the announcement, and ROLM, once again, became the industry leader.

In 1983, ROLM also introduced a family of Digital Desk telephones. At the lower end of the family was single-line digital telephone that included several additional buttons for features. It was designed and priced to compete with the traditional single-line telephones that had been the standard in the industry, requiring a series of codes to access features. It was an instant success and became the de facto single-line set used with the CBX family. The more advanced telephones included speaker phones, multi-line capability, or connection to a PC or data terminal, or they could be configured to complement ROLM's ACD features for call center applications.

There was one other significant occurrence in 1983. Realizing there might be a near-term threat from Northern Telecom or AT&T, both of whom were targeting the office automation market, ROLM management concluded that it would be best to partner with a large computer company. It took only a quick phone call by ROLM to

IBM to confirm their interest in a relationship. IBM and ROLM entered into a confidential agreement to explore the possibility of a significant alliance. Earlier, IBM had entered into an informal agreement with Mitel that would give IBM the ability to market Mitel's new system, SC2000, under the IBM name. By 1983, Mitel had yet to develop the system, and ROLM found IBM eager for an alternative. In June of 1983, IBM bought fifteen percent of ROLM with an option to acquire an additional fifteen percent. Sensing his employees' concern, shortly after the partnership announcement, Oshman sent out a company memo stating what ROLM was *not* going to do. The memo stated that the deal with IBM was not going to change ROLM's philosophy, involve IBM in ROLM's day-to-day operations, or make ROLM "sell-out to IBM." Oshman declared, "ROLM was going to remain independent."

The partnership turned out to be painfully frustrating. IBM had faced an earlier antitrust action and wanted to avoid a reoccurrence. Conducting ROLM/IBM joint sales calls often involved an IBM attorney who might insist on bringing in a major competitor, such as Northern Telecom, to meet with the customer—a most unusual practice in any industry. Development efforts were often thwarted when IBM attorneys insisted that open standards be used so that others could provide product to IBM/ROLM developments. Seeking a solution, Maxfield and Oshman reluctantly became convinced that a complete buy-out of ROLM was the best outcome—which is what IBM wanted in the first place. At the end of fiscal year 1984, ROLM's revenue was a little short of $660 million, with $37 million in profit, eighty ROCO offices, and 9,000 employees—compared to IBM's $46 billion in revenue and 400,000 employees. The purchase was announced September of 1984. The purchase price was $1.25 billion and was approved by the shareholders. The justice department also approved the sale, but required IBM to sell the ROLM Mil-

Spec division, which was a relief to many of the Mil-Spec employees. Loral bought the Mil-Spec group, later selling their entire defense and system integration units to Lockheed Martin in the mid-1990s. In mid-1985, Oshman gave notice he would resign in January. Maxfield was offered Oshman's position as president of ROLM-IBM, which he declined. For personal reasons, he also resigned. He would continue on as needed on a part-time basis. The position of President went to Dennis Paboojian, who once headed operations for Arcata Communications.

The joint ROLM-IBM sales efforts continued to struggle. IBM, instead of embracing the free spirit, hardworking attitude of ROLM, began to insert its own culture. As Katherine Maxfield points out in her book, *Starting Up in Silicon Valley—How ROLM Became a Cultural Icon and Fortune 500 Company*, ROLM employees had no desire to wear blue suits and white shirts. As IBM continued to impose their culture, more ROLM employees resigned. Katherine Maxfield quotes ROLM's head of Service, Bob Finocchio, saying, "How full was the ROLM parking lot at 6:00 Friday night, or on Saturday? Big difference." During ROLM's independence, the lot was partially full almost any day of the week or after hours. Maxfield aptly adds: "The fire in the belly of ROLM was out." ROLM margins were slipping as the industry slowly moved into a slump. Companies like Datapoint, Rockwell, and RCA, who had earlier announced intentions to enter the market, backed out. After the acquisition, during next three years, ROLM-IBM was operating at a loss, and IBM decided to sell the division to Siemens—at the time, a $33 billion company. Siemens had their own PBX, and their engineers argued that their system was superior. The ROLM-IBM 9751 developed during the companies' short marriage won out and was to be Siemens' PBX mainstay for several years—until they introduced Model 30 and Model 80 of their 9006i series, which resembled little of the original

ROLM CBX. Feeling abandoned, many original CBX customers chose to upgrade to competing systems.

• • •

BOOK FOUR

Four Major
ROLM
Distributors

While a few ROLM distributors folded early due to lack of funds, the following four ROLM distributor's stories are well worth telling.

TSA/ATS: Change the Letters Around and Create Another Company, or Two, Three or More

D ave Perdue had sold TSA to Northern Telecom in late 1976, and created a formidable national company for NTI, with sales exceeding $100 million in less than three years. When the new NTI president insisted on moving the headquarters, Perdue resigned NTI in May of 1998 and started over. He re-arranged the company's letters from TSA to ATS—American Telephone Systems—and began selling smaller key systems, obtaining an exclusive distribution agreement from Inter-Tel. Six months later, the company's name was changed to ATS Telephone & Data Systems, as the former name was too close to another company, American Telecom. After securing several contracts and hiring some of his former employees from TSA, Perdue approached ROLM again for a distributorship. Unfortunately, ROLM had appointed another distributor in Memphis and turned Perdue down. Two years later, ROLM had become unhappy with the performance of the distributor for Tennessee and contacted Perdue asking if he was still interested. Of course he was. Perdue told Oshman that he would outspend the existing distributor in marketing the ROLM brand and sell more product. Though initially the ROLM distributorship was only for

Memphis and Nashville, it was later expanded to include Knoxville, and Jackson, Mississippi—much of the distributor area that TSA originally had. Perdue became the only distributor to have held two ROLM distributorships in the same area at different times.

Perdue was sensitive to his customer concerns and objectives and saw an opportunity to provide a long-distance service to them. At the time, there were few long distance carriers other than AT&T, MCI and Sprint, the latter two being the only two dominant competitors to AT&T. Several companies and individuals became sales agents or aggregators re-selling circuits for these few. Perdue took a different tack. He understood the calling patterns of his customers. Most calls from businesses in the South were being placed to Chicago and the East and West coasts. He bought and installed a circuit switch (unlike a PBX which has attendant and station functionality) which, depending upon the number dialed, routed calls over AT&T's outbound WATS lines (800 numbers). Perdue's customers leased access to the service, paying a lower rate per minute than the per-minute rate charged by AT&T. The added value to Perdue's customers significantly reduced the cost of long-distance calls. The company, LD Networks, grew dramatically—much more than he anticipated. Perhaps the rapid growth was a little too much for AT&T. Perdue would form another company, ATS Networks, one more time providing lower cost long-distance service for his customer.

"Again, AT&T tried to shut us down," Perdue recalls. "We went to the Tennessee Utility Commission for a meeting with two of our attorneys. AT&T had a whole room full of attorneys, plus a private line to a hotel across the street with more attorneys listening to the conversation by speakerphone. I was one scared puppy!"

To his relief, AT&T was not successful in thwarting his new long distance business. Perdue's company did so well that a public company purchased ATS Networks in 1982, before his new ATS Networks completed its first year in business.

In 1998, ATS Telephone & Data Systems, started in 1979, and ATS Financial Systems, started 1981, were sold to a utility company headquartered in South Dakota. The utility was buying several Interconnect companies, including another ROLM distributor, KLF, in a "roll-up" effort, with the intent of eventually taking the collection of companies public. Sales of ATS had grown to about $65 million at the time of the sale. Fortunately, Perdue insisted on taking half in cash as the company eventually filed for bankruptcy protection without ever achieving a public offering. Pieces of the company were eventually sold to Avaya. The sale of ATS would not be the last company Perdue would build and sell. He also started ATS Mobile System in 1985—a paging company that was sold to BellSouth in 1988. Ten years after selling the ATS Long Distance company, Perdue and his son-in-law, Tim Whitethorn, started a second long distance company which they sold in 1996 to a forerunner of Qwest. In 2002, Perdue was approached by a software programmer from Bucharest who had helped write software for banks to audit telephone bills. Union Planters Bank was one of the first customers. All product development and programming were done in Bucharest, Romania. Perdue saw the software's value. The Romanian group agreed to contribute the software, and Perdue agreed to finance the company to market it. Perdue would own half the company; the Romanians would own the other half. Perdue was concerned that Union Planters Bank might claim rights to the software, so he asked for a joint meeting of the bank's officers and the Romanian group. The Bank stated they had no desire to be in the software business and would waive any rights they might have to the software in exchange for a fee to be paid to them over the next two years.

"We agreed to that," writes Perdue, "and decided to start the new company in May 2002. After working with an ad agency, we picked the made-up name of Asentinel and started the business. We purposed that all product development would continue by the original software team in Bucharest, but all product sales and support would be handled in the U.S. Asentinel filed a patent application in the winter of 2003 and, even though the patent office declined the original filings by selecting a new patent attorney and modifying the claims, a patent was granted after about five years of trying."

As a starting point, they fed an Asentinel client's telephone equipment inventory record, provided by the local telephone company, into their software. Asentinel employed EDI (Electronic Data Interchange) to automate bill pricing and track all changes for the client. Patents for the unique software were filed in 2006 and 2007. Asentinel software was leased to major corporations such as Apple, Bank of America, Georgia Pacific, and other Fortune 500 companies. In the meantime, Perdue bought out the half ownership of the Romanian group so he would own all of Asentinel except for a position he shared with key employees. The company was sold in 2015. Perdue had a habit of building successful companies and selling them.

CHAPTER 27

UTC/Coradian Corp.: Friend or Foe—How Good Relationships Changed

As many of us did before the FCC ruled that private telephone equipment could be connected to AT&T's network, Stan Ringel and Joe Bruno had a small company in Upstate New York selling intercom and paging systems.

Senator Bruno recalled their first meeting: "I first met Ringel by answering an ad in the paper for a salesman to sell Amana Freezers. Ringel was a broker for Amana, and we sold freezers through a freezer food marketing plan. Every three months we would deliver a package of frozen food to the Amana customer at wholesale prices. The theory was the savings they received would pay for the cost of the freezer. Following that enterprise, we opened Balanced Planning Insurance and Balanced Investors. We sold that company, but a California company who bought us defaulted on the payout. The principal officer of the company was killed in a California mudslide, and the company filed for bankruptcy protection shortly thereafter. We had gone into selling intercom systems by that time."

The company they started was United Telecommunications Corp. (UTC). They sold Ericsson Centrum intercom systems and paging systems in Albany, New York and surrounding area. After the

Carterfone decision, they began to expand their product line, selling Ericsson PBXs and the TeleResource, TR32, which was headquartered nearby.

As former Executive VP, Robert (Bob) Schwartz states, "Bruno and Ringel were the founding partners of UTC. Ringel was full-time, and Bruno was part-time as he was also the chairman of the Republican Party for the county. I had been a consultant for KPMG before joining a client, Gardenway Manufacturing, which made rototillers. When I first met Bruno, I was Gardenway's CFO, and he asked me to have dinner with the two. During the dinner, they asked me to become a Board member for them. Once Joe was elected as New York State Senator, he could only continue part-time as a UTC employee. That's when I joined the company full-time as CFO, COO, [and] EVP. The Board, prior to my joining, was not formal or disciplined. Once I joined in early 1974, I reorganized the company and its Board of Directors."

Ringel was invited to ROLM's headquarters in Cupertino, California, during the same North American Telephone Association convention in San Francisco, when TSI and six other Interconnect companies were also extended separate invitations. "When we saw the ROLM products," Ringel recalls, "we knew we had a winner. Most PBXs at the time were electro-mechanical switching systems. ROLM employed a minicomputer which allowed upgrades, changes, and modifications by changing software rather than hardware." It was a quick decision for their small company.

In late 1974, UTC's sales were under $2 million a year. ROLM gave them exclusive distribution for upper New York State, New York City, and northern New Jersey. After taking on the ROLM CBX, sales began to rapidly increase. Schwartz recalls, "Our first ROLM CBX sale was installed at Central Markets, Inc. [a well-known market chain established in 1933 in Schenectady, New York] which we used as a reference for a lot of other sales." With the ROLM CBX, UTC sales shot up fifty percent to just over $3 million in 1975. "It was a real honeymoon period."

During the first few years, ROLM distributors experienced CBXs crashing all over due to heat, power supply, and software issues. When a system went down, the entire company knew it. "We had major problems [with our first] ROLM systems," Schwartz said. "It was a major PR problem and costly [for us]."

For the fiscal year ending 1977, revenues reached $7.2 million while the demand for ROLM CBXs continued growing. However, UTC struggled to meet ROLM's quotas, which required a fifty percent increase in ROLM purchases each year. In addition, carrying the ROLM product line became capital intensive. ROLM distributors were required to have their technicians who serviced the CBXs carry expensive repair and spare parts kits in their vehicles, and many CBXs in the mid '70s had problems. As long as the ROLM systems continued to lead the market and meet customer demand, dealers, including UTC, were willing to contend with problems inherent in most new systems.

To continue its growth, UTC, like other ROLM distributors, found it needed more capital. In late 1977, they approached ROLM to discuss the possibility of a buy-out. The desired arrangement would allow UTC to continue for another year, after which, ROLM would buy them for a predetermined price. A year earlier in Tennessee, the ROLM distributor TSA had sold to Northern Telecom, a ROLM competitor. Several other ROLM distributors had begun hinting at possible purchases. Ken Oshman, ROLM's President, was "concerned that our distribution would evaporate." The activity was fueled by the distributor's need for capital and the attractiveness of ROLM's escalating stock price. However, their attitude about the purchase of its distributor was less than receptive.

"It was onerous, and we rejected their terms," Schwartz recalled.

Instead, the company decided to raise capital through a private placement of $200,000 in convertible debentures and $300,000 of subordinated debt. Still in need of capital, the following year UTC met with its vendors and banks.

Everyone increased UTC's lines of credit or extended terms—except ROLM. According to Schwartz, "We always sent them our financials each quarter, they knew our financial position, but they wouldn't budge on extending terms of payment. ROLM required payment within thirty days of the shipment, but our customers weren't paying until the system was completely installed. ROLM claimed they couldn't treat us any differently than any other of their distributors."

Also, most all businesses beginning in the late '70s were holding onto their cash as long as possible. Prime lending rates during 1977 bounced around six and half percent to seven percent. By the summer of 1978, the rates were around nine percent, and by the end of 1979, they had zoomed to fifteen and a quarter percent—and were still climbing. In the normal course of business, lenders usually added a point or two to their cost of borrowing. "Cash is King" was the constant refrain.

UTC decided that a public offering would provide the capital base they needed and notified ROLM of their intent. ROLM expressed concern, but said they would not object to the offering. The public offer was led by F.N. Wolf, a small investment house known for promoting under $5.00 a share stocks, often referred to as "penny stocks" traded over the counter. Later, F.N. Wolf would be penalized for "high-pressure sales tactics" and would close in 1984. After the offer in April 1979, UTC encountered a small, interesting problem—but not a bad one. Some investors attempting to purchase the stock of a major independent telephone company, United Telephone, Inc.

(UTI), wound up buying shares in UTC by mistake. The company was sued by UnitedTel, and a settlement was reached. UTI agreed to pay UTC to change their name and agreed to cover all marketing, ad agency fees, and public relations costs. However, the name change had to be completed within six months.

"More than five months had passed," Schwartz recalled, "and we still didn't have a new name. I was watching Star Trek, and Captain Kirk was talking about the planet Coradia, and I thought, 'Hey, that name might work!' And that's how we came up with the name for Coradian."

While talking with potential investment bankers, Schwartz was introduced to a banker who represented Centel Communications Co., a division of one of the country's largest independent telephone companies, Centel. Centel became interested in acquiring Coradian and, as Schwartz relates, "They intended to make a 'very attractive offer.' I thought it would be a win-win for both Coradian and ROLM." ROLM had reserved the right to sell directly to independent telephone companies, so acquisition by Centel could have increased ROLM sales while providing more capital for Coradian to expand. Coradian was again surprised at ROLM's reaction. Instead of embracing the deal, ROLM referred Schwartz and Ringel to a clause in their distribution agreement that allowed ROLM to terminate the agreement in the event of "a substantial change in ownership." Once Centel was made aware of the conversation, they withdrew from further discussions.

It became clear to Coradian that ROLM was controlling their destiny. Going into 1979, Coradian sales would exceed $20 million. However, seventy to ninety percent of their revenue was attributed to ROLM products. The two would have on-again, off-again discussions. ROLM offered to buy Coradian for only "net-book-value," and stated they would not consider any multiple of a distributor's earnings in the transaction. In addition, several distributors who had brief discussions about a possible buy-out were told that the net book value of a distributor's company would be only considered, and that value would be determined only by ROLM after they thoroughly scrubbed the distributor's balance sheet. It didn't matter that many distributors had public accounting firms preparing their annual reports that included what was, in their opinion, the distributor's net book value. The love-hate relationship between the two continued through 1980. For Coradian's fiscal year end, the company's revenue reached more than $36 million. ROLM made another overture which was one-third of what the public was paying for Coradian shares at the time. This was also not accepted. For a while, negotiations stalled. ROLM's third and final offer to Coradian was $13 million at a time when ROLM's stock had declined, $5 million less than what Coradian thought reasonable. Coradian rejected the offer. Not only was it too low but, if accepted, it could have resulted in Coradian shareholder lawsuits.

Both Coradian and ROLM were included in *Inc.* magazine's list of the country's fastest-growing publicly held companies for two years running. Coradian was number fifty-five in 1980, and eighty-three in 1981; ROLM was number twenty-eight and fifty-eight in the same two years. The numbers bespeak an idyllic relationship. The truth, however, was otherwise. Earlier in the year, ROLM had lost another major distributor based in Texas, Fisk Telephone Systems, Inc., which was purchased by Centel. Fisk never had more than fifty percent of its sales related to ROLM, and Centel considered Fisk in a stronger position, unlike Coradian.

Coradian observed more ROLM Operating Companies (ROCOs) being established where distributors once operated. They were fearful of ROLM's ultimate plans in the areas Coradian served. Schwartz and Ringel began doing everything they could think of to improve relations. "We were willing to do almost anything," Schwartz said. "Ten years exclusive agreement, give them a position on our board, issue them a statement [that] we would stay loyal to them. But we never got very far, and that made me suspicious of what their true intentions were."

Coradian began working on a strategy to safeguard themselves from a predatory take over. They began a search of all PBX manufacturers to possibly replace their ROLM products. Other manufacturers had been busy developing software features similar to those offered by the ROLM CBX, and attempted to leap-frog over the minicomputer technology they employed at the time. Coradian was again in need of more capital, and vulnerable. They planned a second public offering through Drexel Burnham and let ROLM know of their intent. ROLM indicated they would not "actively" oppose the offering, but threatened to remove the exclusivity provision in their distributor agreement for New York City.

"Toward the end of the underwriting period and just six days before the offering," Schwartz recalled, "Oshman called Ringel and told him that ROLM was cancelling their distributor agreement [which was due to expire at the end of 1981]. I called Drexel Burnham to let them know of ROLM's decision. Instead of withdrawing, they said, 'Screw them, we'll double the offering.' That pissed off Oshman."

The offering was successful, raising $10.5 million for Coradian, who then began to work with ROLM on terms of a transition agreement. ROLM extended the agreement for another year—but on a nonexclusive basis. That meant Coradian would be competing directly with their principal supplier for a year. ROLM's "National Accounts" program was already fifteen to twenty percent less than Coradian's standard pricing. Given the circumstances, Coradian would likely lose in a competitive situation. The company decided to assign to ROLM the systems that were in progress of being installed, keeping for themselves the installation work and service contracts.

Coradian would eventually decide on distributing a relatively new Wescom PBX. Wescom, a small company based in Chicago, had been in the telecommunication business since 1965, selling T-1 transmission products and an analog PBX to the industry. In 1974, Wescom began developing some early digital switching technology. Sam Pitroda, well-respected engineer and architect of the Wescom 580L PBX, built the system around Intel 8080 microprocessors instead of a minicomputer, which ROLM employed. The company also developed a powerful automatic call distribution (ACD) system that had been purchased by some airlines. However, the development of the 580 PBX and ACD system virtually bankrupted Wescom, with many vendors placing the company on credit hold. By the time Coradian was looking at the 580, Wescom had been purchased by Rockwell International, solving Wescom's cash problems. According to Schwartz, "The Wescom PBX had twice the capacity of the largest ROLM system at the time. It was microprocessor controlled with call detail recording built in, addressed more data applications, and used more efficient cabling [for data]."

Coradian also continued selling Mitel SX 200 PBX and the KeyBX, a hybrid key system that four ROLM distributors and Stan Blau had made for them on an exclusive basis by Telrad, an Israeli subsidiary of Koor Industries. Schwartz would resign from the company in 1981 over a disagreement on the company's direction. Schwartz wanted to lead the company into data management, but the board wanted to stay selling telephone systems.

As Senator Bruno recalls, "In addition to the large Wescom PBX, we started selling the Mitel SX 2000, a system intended to grow to several thousand lines. One sale we made was to a large bank in New York City. Fujitsu saw that we were selling larger systems and became interested in buying us. When Mitel found out, they swooped in and bought the company for $24 million in 1990. They didn't quibble over the price."

Apparently, the Wescom product line was sold to Digital Transmission Inc. who, in turn, sold it to MCI for their network switches and used the ACD platform to sell to public safety agencies, 9-1-1 service.

CHAPTER 28

Jarvis Corp.: A Regional Success with a Local Tragedy

Wilford, "Bill" or "Billy" Anthony Jarvis was known for his outgoing personality and was considered "always a salesman." Jarvis had been working for the advertising firm of Brown & Bigelow and had moved from New York to Richmond, Virginia, in the late '50s to open an office as its divisional manager. Once, on a flight, Bill was reading an inflight magazine and saw an ad for a "continuous play" music device. He tore out the ad and, upon his arrival back in Richmond, ordered the device. He concluded a business could be built with the product—not by selling them, but by renting them. At the time, businesses were paying monthly fees for Muzak background music services in buildings, elevators, and shopping malls. Although the music selections were very limited, the service had become popular in the '60s. Bill decided he could offer a competing product and, with his oldest son, started Jarvis Corp. in the small basement of their home in 1961. A year later, they moved the company into two duplexes. Jarvis also provided paging products and began adding high-end speakers for installations in theaters, music venues, and schools.

As Emmett Jarvis, Bill's son, recalls, "We began beating the Muzak distributor for sales all the time. When the Muzak franchise owner, Walter Cottrell, passed away, Jarvis Corp. wound up with the business. We put sound systems in Scott Stadium at UVA and several high schools, churches, and entertainment events."

The business continued to grow into selling intercom systems—most often competing with the local C&P telephone company. Until the mid- to late-1960s, stand-alone intercom systems used in hospitals, car dealerships, and executive suites were primarily of the "push-to-talk, release-to listen" variety. Ericsson introduced the Centrum intercom systems that offered "hands-free" capability. After dialing a station, the voice channel would automatically switch between two parties depending who was talking. The systems could be installed as an addition to customers' phone systems to improve the distribution of incoming calls, provide paging to specific zones and eliminate the need for a dial-by-station intercom the phone company provided. Jarvis's company was gaining experience in telephony and the tariffs charged by Bell System operating companies. When it became legal to sell and own private telephone systems, selling Ericsson's 741 and 561 crossbar PBXs was a natural extension of their business.

By 1973, Jarvis Corp.'s sales exceeded $1 million. The company was growing fast and moved into a larger remodeled supermarket on Hull Street. Bill Jarvis didn't have to look very far for new employees. The company employed Burt, Billy, Carroll, Mike, Emmett, Betty, Dennis, Joy, Kathy, and Timmy . . . Jarvis. That's right—at one time all ten Jarvis children were employed in the company. Bill Jarvis would bring them into the company as soon they got out of school. The family referred to them as the "upper five and the lower five," indicating the oldest and the youngest. Several of the Jarvis family members would become officers and senior managers running separate divisions.

Michael and Carroll Jarvis, sons three and four, were exceptional football players for UVA. Carroll received recognition as "All State" for Virginia and later "All American" before being drafted by the L.A. Chargers in 1967. Both brothers were tall, good looking, and outgoing. They would eventually become the two principal leaders of the company. Of the two, Carroll was more serious, becoming the company's CFO. Mike had a personality as big as anyone would ever likely encounter. He only had one flaw that I remember—due to his long upper body, he couldn't keep his shirt tails tucked in. Mike was a peripatetic visionary who had the ability to command the attention of those around him. As Emmett fondly remembers his older brother, "He was a real snake charmer—a social butterfly. If we had a problem with the Bell Operating Company or with a customer, we'd send Mike out to resolve the situation. He was the PR face of the company. After just a couple of confrontations with C&P issues, he developed a friendship with Dick Wainwright and Tom Kay, executives at C&P. Wainwright was someone that the company would continue to rely on to resolve telephone company related issues."

Emmett started with the company at age fifteen, eventually becoming responsible for what he calls the "back-end" of the business—working with Billy dispatching and managing the installation and service portion. Emmett remembers: "We also sold the Northern Telecom SL-1 and their 'Pulse' PBXs. The Pulse was our best-selling PBX—we had absolutely zero problems with them. Our biggest problems were with the protective couplers [PCAs] that Bell made us connect to. They would reduce the volume and created nothing but problems. We first complained to C&P, but when they finally got back to us, they claimed it was our problem. We started telling our installer to wire around the PCAs [directly connecting into the Bell network]. The problems went away. It became routine to bypass them as part of our standard installation practice."

Jarvis Corp. would continue growing, expanding by opening offices in Atlanta, Georgia, Baltimore, Maryland, Washington D.C.,

Charleston, Norfolk, and Savanna, Georgia. In 1977, Jarvis Sr.,
decided to retire and stepped aside, allowing his son, Mike, to run
the company as CEO. Jarvis, Sr. moved to Charleston to retire. But
retirement didn't suit him, and he elected to start a long distance
resale company, which also became successful.

J. Michael Jarvis, Richmond New Leader, Feb. 19, 1979

Jarvis Corp. began selling larger systems. A Northern Telecom SL-1 was sold to the Federal Reserve Bank in Richmond. After picking up the distributorship for the ROLM CBX, the company landed a large 8,000-line contract for the University of Virginia in 1979. The sale was significant in many respects. It would become a springboard for larger commercial sales and government agencies. It became the largest ROLM installation at the time and required a lot of support from ROLM, which they willingly provided. During the installation, the relationship between the two companies also grew.

Jim Bloom, who had been working for IBM before being hired by Jarvis Corp, remembers: "It was an entrepreneurial experience! At first, I thought I had made a mistake. I was coming from a very structured environment where one was expected to maintain the IBM image, even the dress code and attitude—you couldn't even drink [during working hours]. At Jarvis, everything was unstructured."

Nick Visser had known Michael and also had a friend working for Jarvis who suggested setting up an interview. Visser had been a salesman for Dictaphone selling answering devices when Jarvis recruited him in 1972. Visser credits Bloom for playing a large role in Jarvis Corp.'s growth. "Bloom drew up marketing plans, refined proposal development, established pricing guidelines, set quotas, and added structure to the company," he said. Bloom would continue to assume significant roles with the company and its successors.

In 1976, having proved himself a successful salesman, the company offered Visser an opportunity to open a branch in Baltimore. Starting with only himself, Visser grew the branch with added salespeople and technicians, and expanded into the D.C. area. Outgrowing two other locations, he established an office off the 295 Parkway between Baltimore and D.C.

In June of 1980, ROLM reached an agreement to purchase their portion of Jarvis Corp's business. During the early '70s, the company established a finance division, Charter Leasing, to help finance the sale of systems the company sold. After ROLM's purchase, the smaller key system portion of the business, leasing, music, paging, and basically everything not related to ROLM was folded into the remaining Jarvis Corp. Mike Jarvis would continue to head up the ROLM Atlantic division, and Carroll was made President of the new Jarvis Corp.

The celebration of selling part of the company and Carroll's promotion was short lived. Eight months after the purchase of part of the company by ROLM, Carroll was driving his new Porsche when he hit a bridge abutment, the car landing in a creek below. He managed to extricate himself, but fainted, falling into the creek where he drowned.

After his brother's death, Mike resigned from ROLM and assumed the position of Chairman of the Board for Jarvis Corp. Jim Bloom also resigned from ROLM to become VP and General Manager of Jarvis Corp. Visser remained with ROLM, becoming VP and General Manager of ROLM Mid-Atlantic until November of 1983, when he also resigned to become President of Jarvis Corp., a position he would hold until 1986.

The period of mid- to late-1983 was a pivotal year for many. Mid-year, IBM bought part of ROLM with an option to buy more. During that announcement and IBM's total acquisition of ROLM the following year, Kessman, Mike Jarvis and Shlomo Shur, a key engineer from TIE, formed ISOETEC to manufacture high featured hybrid key systems. Following the successful introduction of ISOETEC products, Jarvis Corp. began to sell more ISOETEC systems than the TIE products which had been the company's staple since the early 1970s. In early 1986, Jarvis Corp. was purchased by ISOETEC. In October the following year, ISOETEC issued its first public offering.

CHAPTER 29

Norstan, Inc.: How to Build a Large Company–Start with a Financial Background

P aul Baszucki grew up in Saskatchewan, graduating from the Canadian School of Business with a degree in finance and accounting. He taught accounting for a short while before becoming a Chartered Accountant, the Canadian equivalent to a CPA, and was hired by an accounting firm to conduct audits. His next step was working for IBM in Canada where he was an analyst. In 1971, Baszucki went to a recruiting firm that set up an interview for him with a new company, Arcata Communications. Just before the interview, he was told by the recruiter that they already hired someone. He went to the interview anyway, presenting his credentials to John Barbour, then Vice President of Arcata Communications' Chicago branch. Baszucki made a clear argument: "You hired the wrong guy! I was the right guy for the job!" Barbour agreed, withdrawing the offer to the other applicant and hiring Baszucki on the spot.

A year earlier, Barbour and his partner Bud Toly had a small communications company, National Communications Planning Service (NCPS), in San Jose, California. They had installed a communication system for a San Jose customer who wanted the same system installed in the office of their Chicago branch. When the two discovered how much higher Illinois Bell rates were for business telephone equipment as compared to the rates charged by PacBell for the same equipment, they packed their bags and left their small company in San Jose and headed to Chicago to start another company that was also acquired by Arcata Communications. In the early 1970s, a few firms that wanted to enter the business on a national scale decided they needed to move quickly—often throwing caution to the wind and failing to provide adequate due diligence in their hiring practices and acquisitions.

"The books were a mess. It was a wild place, and a lot of cleanup had to be done," remembers Baszucki about his experience in Chicago. Within a year of Baszucki assuming his position as comptroller for the Chicago branch, Barbour was transferred back to California to become Regional Manager of the San Francisco office, and Baszucki was let go. Arcata National had sold Arcata Communications to General Dynamics at a loss of about $25 million and the company was put under their subsidiary, Stromberg-Carlson. Most of the Arcata/Stromberg-Carlson offices were eventually closed.

When Barbour left Arcata Communication to assume a VP position with Litton Business Telephone Systems (LBTS), he asked Baszucki if he was interested in becoming the LBTS Minneapolis Branch Manager. Now having some sales experience, Baszucki accepted and commuted between Chicago and Minneapolis until moving there late in the fall of '73. By the end of the year, Litton BTS had begun closing down their offices.

"Here I was working for two large companies that went bust," Baszucki recounts. "They didn't know how to run a business. I have to get into (the Interconnect) business myself. When Litton closed the office, I told some prospective customers I had been working with, 'I'll be back.' I called a friend, Phil Salvatori, who was working for AllCom in Chicago. 'I need a job.' Salvatori put me in touch with Sid Cohen. After a couple of meetings with Cohen, we decided there was an opportunity before us, and we had to move quickly to take advantage of it. Going into business with Cohen was done by a handshake in December of '73."

Cohen knew of a small manufacturing company, Norstan, Inc. that had gone public to raise cash but was not growing. The two reached an agreement with Norstan and started their sales activities; securing contracts for three TR/32 PBX contracts in the first sixty days.

"We bought LBTS's inventory in Minneapolis for five cents on the dollar and began servicing the LBTS customer base as well as some Arcata customers. We also bought the inventory of another competitor whose business had been taken over by their bank. Now, we had a decent-sized customer base, inventory, new sales, and we were a public company. The fact that our financial information was publicly available gave us a definite advantage over our competitors who were private," Baszucki adds.

During the first two years, concentrating on only the local Minneapolis area, Norstan mostly sold TIE key systems and TeleResources TR32 PBXs. As Baszucki recalls, "We couldn't make much money with TIE [local competing firms were offering the same TIE systems, eroding margins] and key system sales didn't drive revenue much. In 1975, ROLM appointed us a distributor. After that, we never looked back." Norstan also had technical problems with the initial ROLM CBXs. Regardless, Baszucki decided they were not going to sell any products that competed with the CBX. In 1976, Norstan issued a second public offering.

"We rapidly became ROLM's biggest distributor in the Midwest," Baszucki explains, "expanding into Ohio and Kentucky by acquiring Electronic Engineering Company [EECO] in 1985, another ROLM distributor owned by Marty Liebowitz. Also in '85, we acquired Larry Bushkin, the ROLM distributor for Arizona and New Mexico."

Norstan grew rapidly from a local Interconnect company to a strong regional and eventually an international one, becoming ROLM's largest independent distributor.

As the industry matured, AT&T had developed "competitive squads" in some of their regional operating companies to confront the growing competition. Baszucki concluded Norstan could benefit from the experience of one of these "squads" and decided to create his own by hiring one of Northwestern Bell's squad leaders who brought with him three or four people. The increasing size of the ROLM product line and knowledgeable former AT&T marketing personnel provided Norstan the confidence to win bigger sales and tackle large installations previously only dealt with by AT&T's operating companies.

"One example was a ROLM sale we made to the state of Iowa of 15,000 to 20,000 lines," Baszucki said. "Our forte was that we just nailed our customer service record. When we got into a competitive situation with Bell, we challenged the prospective customer to call Bell customers for a reference, then call ours. That usually made the sale. We primarily became a ROLM company, resulting in less inventory we had to carry."

Sensitive to new products coming to the market, Baszucki was not content with just the sale of telephone systems. He moved Norstan into additional products and services that would benefit his growing telephone customer base. By the early '80s, all Interconnect distributors had realized the value of their customer bases. Customer moves, additions, and changes to their system often resulted in greater profitability and each year contributed to a larger percentage of the company's total revenue. The easier and less capital-intensive path to growth for many in the business was to remain local and distribute ancillary products such as voicemail, interactive voice response systems, and call detail recording units that could be added to most PBXs and Key Systems. While Baszucki did sign distribution agreements with some voicemail and voice processing products, he had a much larger vision, expanding geographically while providing long-distance service and integrating voice, data, management, and consulting services through a series of acquisitions.

Baszucki proudly states: "Norstan had a great people culture. We constantly measured employee satisfaction as well as our customers satisfaction." By the late '80s, Norstan's revenues had grown beyond $100 million a year with modest but consistent profits. In July 1990, they entered into an agreement with ROLM to refurbish and resell used ROLM CBX systems and components. As ROLM espoused a philosophy of "backwards compatibility" and "upgradability," the partnership benefited both Norstan and ROLM sales forces by providing another option to offer prospective customers. The agreement called for a sharing of profits and allowed Norstan to provide installation wiring for the CBX systems that ROLM sold directly. Later, Norstan would also refurbish and resell Northern Telecom equipment. In April 1992, they acquired the Canadian assets of ROLM/IBM after IBM's acquisition. Norstan's new Canadian segment would incur losses in '93 and '94, but obtain a profit in '95, as Baszucki reduced the Canadian operations' overheads.

Between 1991 and 1995, revenues more than doubled, increasing from $134 million to $290 million with net profit of $7 million or 2.4% of sales, the profit percentage consistent with previous years. Revenues continued to increase to $456.4 million in 1998—the highest the company would achieve. However, profits as a percentage of revenues declined to $3.8 million, or less than one percent. By the late '90s, the company's revenues were almost equally split between telecommunication-related products and information technology products and services. Norstan's "Communication Solutions" division consisted of the design, sale, and installation of telephone systems, call centers, voice processing, service, and moves, adds, and changes to customer systems. Norstan's "Global Services" group included information technology (IT), developing internet solutions, data applications and integration, consulting services, technical education and training, as well as managing customer networks. "The IT service business helped drive revenue in the mid- to late-'90s, but by '99 and '00, that market crashed," explains Baszucki. Although well diversified to meet customer needs, the early years of 2000 were a challenge for the company as sales began to decrease with losses incurred in '00 and '01 which were largely attributed to discontinued operations and disposal of those operations.

The American Arbitration Board on February 25, 2002, awarded Norstan $7.2 million from PRIMA, a consulting firm that Baszucki had purchased earlier. Baszucki stated, "The president, Michael Vadini, had violated his noncompete agreement by going back into business, stealing Norstan customers." As recorded in the company's annual report on April 30, 2003, Norstan had offices in twenty-two cities, five of which were in Canada, and a total of 1,300 employees.

In January of 2005, Norstan was purchased by Black Box Corp. for $82.3 million. Baszucki recalled revenue at the time was running between $250 million to $300 million a year. That revenue estimate was possible even though Norstan's revenues for the trailing six months ending October 30, 2004, were only $111 million with profit of $2.6 million. Profits for the six months were up more than $7 million from the '03 loss of $5.4 million during the first six months of Fiscal Year '03. However, Norstan's revenues had been steadily declining.

For five years prior to purchasing Norstan, Black Box's revenues had been declining from $827 million in sales in '01 to $536 million in '05. However, each year, the company had been able to demonstrate reasonable net profits of between six to nine percent. Black Box saw a merger of the two companies as a definite win, providing comple-mentary services cross-selling into each other's customer bases. Black Box would continue on an aggressive buying spree. Some of the new "Baby Bell" companies resulting from the divesture of AT&T had purchased Interconnect distributors but found the enterprising organizations were difficult to manage and sales were eating into their former regulated business. Black Box took advantage of the Baby Bells' disappointment, and in '08 purchased B&C Telephone, Bell South (dba AT&T Southeast Communications), and an AT&T/NEC TDM business. Their revenues increased in fiscal year ending (FYE) 2006 to $721.3 million. From 2007 until 2011, revenues remained relatively flat at $1 billion while achieving consistent net profits. Everything looked like it was going according to plan. Ending fiscal year 2012, revenues remained constant, but the company incurred a loss of $246 million. Black Box's stock in 2011 reached a high of $40.78 a share with a low of $26.40. For the next four years the company's stock would bounce between $19.80 to $36.20 a share while the company displayed obvious signs of stress. Sales continued to flirt at the $1 billion mark, but net income began to falter. A slim profit was obtained in FYE 2013, but net profits for '14, '15 and '16 were a loss of $116 million, net profit of $15.3 million, and a loss of $171 million, respectively. Black Box lenders had become very nervous and wanted the company to refinance or find other lenders to take them out. Sales began to decline in '17 and '18, reporting additional losses of $7 million and $100 million for the two years. On December 24, 2018, Black Box sent a letter to all their share-holders stating that the company had "run out of time with its lenders and was unable to locate alternate lenders." On January 7, 2019, AGC Networks tendered an offer to buy Black Box for $1.10 a share, or $17.2 million. The offer was accepted by shareholders a few days later. Ten days later, the company suspended trading.

• • •

BOOK FIVE

The Industry Grows

CHAPTER 30

Inter-Tel Inc., Part II

"When You're Going Through Hell, Keep on Going."
—Sir Winston Churchill, and favorite saying of Steve Mihaylo

S teve Mihaylo had persisted in starting from nothing to building a national company. He had worked through the typical startup problems and eliminated a $2.5 million inventory of the Key-Lux, not the most sought-after phone system in the market. After three years of development, Mihaylo and his team introduced a new product, the SPK, with advanced features and increased capacity. The Inter-Tel dealers clamored for it.

In early 1981, the tide had turned in Inter-Tel's favor, but there was a huge problem building. The new SPK system upon which Inter-Tel's future rested didn't work. In any business, management expected occasional problems with a computer, the AC system, or a copier. The telephone system was expected to be 100% reliable, as it was the life blood of the business. If customers could not reach one company, they could dial another number and shift their business to a competitor. SPK customers began to experience numerous problems of dropped calls or calls that would fade away or merge with another call. The worst problem was when callers would tell the SPK customer that they had been trying to reach the company, but all they heard was "ring-no-answer." Inter-Tel was inundated with a flood of service calls. Customers began to threaten lawsuits, occasionally saying, "we're going to throw the system in the street and go back to Bell."

The problem was that the software programmers had virtually no prior experience in writing code for the new microprocessors. At first, the Inter-Tel software engineers would isolate the problem and write a software patch to fix it. But the "fixes" often created other problems. Undaunted, Mihaylo and Sherman knew something had be done. And they knew the solution was not likely to be found in-house. In August of 1982, they recruited two engineers, Ed Terminy and Gehardt Klaiber, who both had experience writing code for the Siemens 192 PBX. Klaiber had left Siemens earlier and talked Terminy into taking on the SPK problem. Terminy recruited two of his best engineers. The SPK lacked basic system protection, so the first task was writing "recovery logic" or a soft restart so that a minor failure would not bring down calls in progress. While the initial recovery logic quelled the swelling wave of complaints, Terminy found the basic software was "spaghetti code"—just code piled on top of code. A total rewrite of the basic software program was necessary.

During this period of SPK troubles, Mihaylo recognized that he needed more senior management at Inter-Tel and was persuaded by some of the company's Wall Street underwriters to get "someone with gray hair to help run the business." Sherman, the current EVP, had done a great job building the distribution network, selling some $11 million worth of product at eight percent commission. But Mihaylo knew Sherman was emotional when it came to facing problems, preferring to avoid them when possible. Senior management would help stabilize the staff and allow Mihaylo to address issues and tasks outside the office. Mihaylo approached Richard Long, former VP for Fisk Electric in Texas, who was then President of the North American Telephone Association. NATA's membership now included telephone manufacturers, importers, distributors, suppliers, and Interconnect contractors. Long was offered and accepted the role of executive vice president.

Since the beginning of the SPK problems, Sherman had been unhappy. He left shortly after Long arrived, coaxing some key Inter-Tel personnel to go with him and start a new company, Vodavi (an amalgamation of Voice, Data, and Video) which would compete with Inter-Tel. The thought of a key executive leaving, weakening the company by taking key employees, and forming a competing company was completely foreign to Mihaylo's concept of ethical business conduct. He also learned that, while visiting Taiko in Japan, Sherman had met with Lucky Goldstar (LG), a large company closely controlled by a wealthy Korean family. LG was interested in becoming a supplier to Inter-Tel, but Sherman steered Mihaylo away from the relationship. LG later provided financing for Sherman's new company. Mihaylo sued Sherman for breaching Inter-Tel's non-compete agreement, and although the suit was later settled, the relationship between the two friends was forever destroyed.

Long moved into the position of president and COO, but he and Mihaylo disagreed on the course the company should take. Two years after accepting the position, Long resigned and returned to resume his role in NATA. Fraught with the turnover and SPK problems, Inter-Tel posted a loss of $1.8 million in 1983. The stock price had fallen to $5 a share, and its reputation in the market suffered. Inter-Tel would experience a loss of $2.4 million in 1984.

Mihaylo knew that a reduction in staff was necessary to save the company—the alternate to Long's preference of increasing sales. Between 1983 and 1984, he reduced the staff from 480 employees down to 260—a forty-six percent reduction. It was a very painful process for Mihaylo, who had instilled faith and loyalty in his employees. While the SPK problems continued to plague the company, Mihaylo met Maurice Esperseth, who was in charge of GTE's research and development facility in Phoenix. After working thirty-three years for GTE, Esperseth had become disenchanted with the large bureaucratic company. When he shared his feelings, Mihaylo promptly offered him the role of head of research and development at Inter-Tel. He accepted the challenge. Esperseth contacted Dr. David Pheanis, a professor of engineering at Arizona State University, and asked him to come over and take a look at the SPK problems. Pheanis quickly concluded that the hodgepodge of SPK software was not repairable, that new software with sound structure would need to be developed, and that Inter-Tel's software department lacked the expertise to write it. With Mihaylo's support, Pheanis selected a team of eighteen engineers, seventeen of whom had been top students at ASU. Pheanis knew their methodologies would be similar. For the next eighteen months, the team worked on the development of Inter-Tel's second microprocessor-controlled key system, called the GX-120. They worked in a separate facility to avoid friction from the existing SPK engineers—most of whom were hardware engineers.

After almost two years of work, Inter-Tel announced its new system, Galaxy or GX-120, in 1985. The GX-120 employed analog phones controlled by a digitally switched microprocessor. The Galaxy contained advanced features normally found in very sophisticated PBXs like ROLM. The telephones included liquid crystal displays capable of displaying information on the status of internal personnel and calls in progress. In 1986, Inter-Tel's relationship with Taiko had become rancorous, leading to a $100 million lawsuit filed by Inter-Tel against its supplier of fifteen years. Commenting on the suit, Mihaylo stated, "They ripped us off, stole our design, and started selling Inter-Tel-like products in the U.S. and Japan." Mihaylo, in turn, entered into a multimillion dollar agreement with Samsung for the manufacture of Inter-Tel's phone systems.

Up to the introduction of the GX 120, no other key system or small PBX offered GX's capabilities at a reasonable price. The GX would become the company's most successful product and restore its reputation. However, the company still incurred a loss of $830,000 for the year. TIE, Inter-Tel's largest competitor in the under-100 station market, was about ten times Inter-Tel's size and had begun to flounder. TIE had flooded the market, and dealers were becoming disillusioned with declining margins. Mihaylo estimated, "At the time of a down economy TIE probably sat on a year's worth of inventory, where we had only a month of inventory."

With a highly marketable product in the GX, during the next two years, Mihaylo again began to expand, acquiring dealers. He negotiated a private label agreement with North Supply and provided them a proprietary change in the telephone case which was called "Premier." After so many years of struggling with the SPK, Inter-Tel again returned to profitability, earning $400,000 in net profit in 1986. In 1987, sales rose to $50 million, and earnings jumped to $2.2 million. Just as the company seemed poised to achieve the sales and profitability Mihaylo had anticipated, AT&T filed a complaint with the U.S. Department of Commerce, claiming that foreign manufacturers were "dumping" products into the U.S. below their actual cost. This led to a 13.4% duty penalty on Inter-Tel's Korean subcontractors and resulted in a decline in profits for 1988.

By the late 1980s, Inter-Tel introduced expanded versions of the GX 120, the GX 400 and the Premier (sold by North Supply) ESP DX. The two could handle up to 832 ports (used for both trunks and stations), pushing Inter-Tel into the market dominated by Northern Telecom, ROLM, and NEC. These companies were producing all digital products by the late 1980s, whereas the Inter-Tel GX 120 and GX-400 still employed analog switching technology. Inter-Tel's newest product was being eclipsed by advances in technology. The product line that saved the company in the last half of the '80s was now obsolete.

To create a place in the merging telephone and computer desktop market, Mihaylo understood that the company would need to replace its GX product line with an all-new all-digital system. Development on the AXXESS began in late 1988. North Supply had tired of the GX. They wanted an all-digital product and didn't want to wait. By mutual consent, the agreement with North Supply was not renewed in April of 1993. In the meantime, Mihaylo continued with his acquisition of dealers. In 1991, Inter-Tel had about 600 employees, including their eighteen direct sales offices. Inter-Tel also established NetSolutions, a long distance resale division offering reduced long-distance calling for its customers. However, the division experienced little growth as it wasn't supported by the sales field. Although the company expanded in 1991, it also incurred a loss of $4.2 million when it had to write off a $6 million hotel and office complex that the company had earlier pledged as security for a loan. In 1992, the company achieved record sales of $79.4 million. It had taken five years to develop the AXXESS which was finally introduced in December of 1993. The AXXESS employed digital signal processors (DSPs) which were much faster than earlier microprocessors. It supported computer/telephone integration (CTI) providing features that were only available on much larger systems at the time.

By 1996, Inter-Tel had twenty-five direct sales offices, more than 1,000 dealers, and an additional twenty dealers in Europe. AXXESS 3.0 increased capacity to 256 ports, and the company also introduced AXXENT, a low-end product that addressed the under-sixteen telephone market. Inter-Tel engineers had also been following what was happening in the development of the internet. Email was becoming a staple for most companies shifting messages inside and outside the company. It would only be a matter of time before voice, which could be digitized, would be handled the same way. In October of 1997, Inter-Tel released its internet telephone product,

VocalNet, which employed Voice over Internet Protocol (VoIP).

Inter-Tel was ahead of the market for a change. Sales for the year grew to $223 million, and net income increased to $14.7 million. Investors also felt Inter-Tel was on the right track and sucked up another three million shares of Inter-Tel stock at $21 per share. Inter-Tel's NetSolutions had been steadily growing, but it was still reselling time from other providers such as MCI and Sprint. Mihaylo believed that the company could establish its own internet long-distance service, attracting a lot more customers and obviating the need to purchase wholesale long-distance circuits. Inter-Tel quickly signed on several international partners that would connect into Inter-Tel's IP network. Most all switched voice network providers, including AT&T, were reluctant to embrace the VoIP technology as it would eat into their legacy circuit switched networks. At the time, several tech gurus were predicting the demise of voice over the internet, as the increase of traffic would diminish the quality of service. Mihaylo was betting on an opposite outcome.

The following year, 1998, Inter-Tel made a quantum leap in two directions: AXXESS 5.0 was released, expanding the system from 2,000 to 3,000 phones, placing it in competition with the larger players like Nortel, NEC, and Siemens (formerly ROLM/IBM). The company also acquired Telecom Multimedia Systems, Inc. (TMSI) for $25 million. The latter brought valuable voice compression software, packet data, and routing capability which enhanced Inter-Tel's Voice over IP capability while reducing costs to outside vendors. By the end of 1998, the company had only $6 million in internet-related services. That would soon change. During the year, Inter-Tel signed on a Japanese partner, and twelve additional European cities had also signed up for voice over internet service. From the early 1990s on, the company demonstrated considerable growth in sales and profits, ending the year with reported sales of $274.5 million, with $22.7 million in net profit. Inter-Tel had started paying dividends the previous year.

In 1999, Inter-Tel acquired Executone's sales and distribution network for $44.3 million after Bain tried to merge Executone into Staples. At the time, their distributor network revenue volume was running just under $200 million year (about $100 million wholesale). Executone had gone through numerous owners, failing to generate any significant profit. Previous agreements between Executone and their dealers were included in the sale. Inter-Tel saw the Executone dealer network as an augmentation to their existing network through which Inter-Tel products would be sold.

However, in several locations there was an existing Inter-Tel direct office and a competitor selling their systems. Inter-Tel sent a letter to all dealers assuring them of the continuation of availability and support of existing Executone products. To add value to Inter-Tel, the dealers would need to sell Inter-Tel's products—that was the rub. Several dealers felt betrayed, claiming Inter-Tel failed to live up to support and availability for Executone products. Six dealers wound up filing suit, pleading injury. Mihaylo recalls only the one suit filed by Executone of Columbus, stating simply, "We lost." However, in 2005, Inter-Tel lost a $7.8 million "breach of contract" claim filed by Executone of Florida. The judgment was offset by an award of $435,000, resulting from a promissory note to Inter-Tel owed by the Florida dealer. Inter-Tel wound up writing off the entire acquisition cost in year 2000, incurring an after-tax loss of $29 million for the year.

Steve Mihaylo (R) with Inter-Tel distributor, Barry Wichansky, circa 1995.

Inter-Tel's success continued into the mid 2000s with sales ranging between $380 million to $450 million and an average seven percent after-tax net profits. Everything seemed to be flowing in Mihaylo's favor—except for a couple of major hiccups, one perhaps leading to the other. In the spring of 2004, with a little encouragement from Anil Puri, Dean of Cal State Fullerton's school of business, Steve Mihaylo pledged $3 million to CSUF, his Alma Mater. It was the largest pledge ever received by the university at the time. In return, the university agreed to place Mihaylo's name on a new business school building. Elated about the pledge, CSUF professor Hamid Tavakolian tasked his students to research Mihaylo so that others could learn what it takes to become such an outstanding graduate and role model. The students' research came back with some surprising information—information the university administrators found very disturbing.

... AND HERE COME THE CONSULTANTS AGAIN

From 1999 to 2001, Earl Nelson, Inter-Tel's Branch manager at the time, had responsibility for sales and service in Bay Area. One of his salespeople and top salesman for Inter-Tel, Bill Bryant, was working with Judy Green, a consultant who was designing an internet access system for the San Francisco Unified School District under a government funded "E-Rate" program. The E-Rate program provided discounts to eligible schools and libraries to obtain affordable internet services depending upon the district's poverty level and status. At some point, the consultant and Brant appeared to agree that the consultant should also be paid a fee from Inter-Tel. The proposed contract may have been inflated to cover the consultants' requested fee. Regardless of how it was arranged, the end result was that the school district would have awarded a contract that involved some type of "kick-back." The SFUSD became suspicious and filed suit for bid-rigging in May of 2002, naming several entities involved, including Inter-Tel.

Federal prosecutors took over the lawsuit and spent two years investigating the case. Ultimately, six individuals and six companies were indicted by a grand jury. Inter-Tel was named as one of the six companies along with Earl Nelson. Mihaylo was not indicted, but Inter-Tel pled guilty, and in January of 2005, paid $1.7 million in fines plus an additional $7 million in restitution. As Nelson remembers, "Hell, I didn't have anything to do with it, didn't know the consultant or contacts at SFUSD. We just had these two FBI 'Nimrods' poking around looking for a target. My attorney wanted me to take a plea-bargain, but I said 'No,' not when I didn't do anything."

Mihaylo supports Nelson, saying that, "Nelson was a good guy. The real culprits were his salesman and the consultant." In the same scheme, NEC paid a $20 million fine. Later in the year, after Inter-Tel was purchased, Nelson was terminated.

Mihaylo increased his CSUF pledge to $4.5 million, and any issue concerning his pledge to the university evaporated. The following February of 2006, Inter-Tel's board chairman, Alexander Cappello, requested Mihaylo resign or face a vote on terminating his employment. Mihaylo resigned and was replaced by Norman Stout as CEO. Stout launched a takeover bid. Mihaylo later claimed his ousting had no connection to the SFUSD fines but that the board simply wanted to move in a new strategic direction. Mihaylo fought to rejoin the company, waging three proxy fights and winning the first one. The second resulted in a stalemate, and Mihaylo lost the last with shareholders narrowly rejecting his bid.

Inter-Tel's board, as Mihaylo related, had begun a search for a possible purchaser of the company behind his back. The board started discussions with Mitel Networks, which had been planning an initial public offering. While flirtations between the companies continued, Mihaylo, teaming with Vector Capital, submitted an offer at $23.25 a share to purchase Inter-Tel. But the offer was rejected by the Board. Vector promised to return with another offer between $26.50 to $27.00 a share. The board seemed to want to avoid another proxy fight plus a looming Mitel termination fee of $20 million. They accepted Mitel's increased offer of $25.60 a share. Inter-Tel's estimated revenue at the time was projected to be $388.7 million with pretax profit of $45 million. In August of 2007, Inter-Tel was sold to Mitel Networks for approximately $729 million—all cash. While Mihaylo would have preferred to continue building Inter-Tel, he had achieved his goal of building a profitable company that was sold for cash and stated proudly that he had returned over $1 billion to shareholders.

At the time of the sale, Inter-Tel had 2,500 employees, had amassed 1,500 patents, and paid no royalties. Regarding any problems with AT&T, Mihaylo stated, "For twenty-five years, ATT/Lucent would send in their attorneys claiming we were infringing on their patents. They would spend a day or two going over them and then take off and play golf the rest of the week. Most of the time, they were the same attorneys. . . . They never found anything. We offered a mutual cross-license deal that would have provided Inter-Tel to AT&T's vast patents, but the offer was never accepted."

And how did Inter-Tel fare after the acquisition by Mitel? Did it share a similar fate as ROLM after being acquired by IBM, and later, Siemens? One can look at Mitel's sales and earnings prior to acquiring Inter-Tel. As a Mitel annual report at the time states, "We have incurred net losses since our incorporation in 2001 and may not be profitable in the future."

Not exactly what an investor wants to hear. Mitel's revenues for the past several years had been similar to Inter-Tel's but without any earnings. Mitel's accumulated deficit at the end of April 2007 stood at just under $400 million. After the acquisition, Mitel reported sales of $692 million, and the company appeared to have absorbed and capitalized on Inter-Tel's products and services. However, losses continued until 2010, when Mitel reported its first profits of $37 million. Mihaylo indicated his disdain about the acquisition, stating, "We paid $3 million for a software license from Cilantro and invested $100 million in it developing our cloud-based services. Due to Terry Matthews' 'Not-Invented-Here' attitude, Mitel trashed it."

Sales for Mitel remained flat for the next two years, but profits began to decline in 2013. In 2014, Mitel reported $1.1 billion in revenue but a loss of $7 million. Until 2018, revenue remained around the $1 billion mark, but losses continued. The company's earnings before interest, taxes, depreciation, and amortization (EBITDA) showed that they had the ability to earn profits once loans were repaid, write-offs of the premiums paid on acquisitions completed, and other adjustments made. The stock price between January 2016 and May of 2018 ranged between $6.00 to $11.30 a share. In a December 10, 2018, SEC filing, the company announced completion of an acquisition by MLN, UK, Hold Co. LTD. The all-cash offer to Inter-Tel's shareholders seemed to be in their best interest. What cannot be determined is if they would have benefited more had Inter-Tel continued as an independent company, run by Mihaylo.

CHAPTER 31

Vodavi Communication Systems, Inc.: A Quintessential Entrepreneur

I t was cold and depressing in East Flatbush, Brooklyn, where Steven Sherman grew up.

"I grew up pretty angry in a poor neighborhood with a gloomy atmosphere," he recalls. "My parents were immigrants and sick a lot. Both went into the hospital for a year. My baby sister was born when I was sixteen, and I remember coming home from school to take care of her. If I wanted to go somewhere, I had to take a bus."

Since he provided care for his family, Sherman was classified as "head of the household" by the local draft board, reclassifying him from 1A to 3A. This meant a draft deferment. Sherman attended New York City College, graduating from the Baruch School of Business in 1968. After college, he enrolled in a six-week ITT course to obtain a teacher's credential. A teacher's license would extend his draft deferment. He only taught class once. During the summer, Sherman went to an employment agency to secure a job until the fall semester. The agency set him up with an interview with New York Telephone Company (NYTel), and he was hired. NYTel found him to be bright with an effervescent personality, and soon placed him in their Junior Executive Training Program. Becoming a teacher was shoved aside. After one and a half years with NYTel, he was hired by Phone Consultants of New York, owned by Art Gallob. The company competed with NYTel, doing audits and selling intercom systems, using the "5 Step Program" similar to Craig Dorsey's company, CCI, in San Francisco.

"Art's company was located in a dingy storefront—also in Brooklyn—with offices upstairs. Walking on the sidewalk, one could easily mistake it for a dry-cleaner's shop. I was still taking the bus to work. This was about a year before Arcata bought Gallob's company."

Frank DeFoe, who had been Arcata Communications' Denver branch manager, came to the New York branch for a short while. It is unclear what DeFoe's official role was, but Sherman credits him with changing his perspective: "DeFoe made an impact on me as he changed how I thought about my life—I was about twenty-five at the time, and he was older, fit, more upscale. He decided that our office in Brooklyn was not up to what Arcata should have, and leased a new location at 90 Park Avenue in Manhattan."

Gallob was eventually replaced by Darrel Nelsen who had been tasked to open Arcata's Communications office in Boston. "It was a great location, across from the PanAm building," Nelsen recalls, "but we couldn't make any money due to the high overhead. Maybe we made a little, but we had twice the number of CWA installers than we needed, and we were locked into a long-term lease. I cut the installation force in half during the first three months."

Sherman attended an Arcata Sales conference in Glen Cove, Long Island, where he met Tom Kelly and Gil Engles from TIE who were invited to give a presentation of the products they were importing from Japan. It was also my first meeting with Sherman.

After Arcata, Sherman joined his friend, Stan Blau who was working for Plessey at the time. Blau had a childhood friend, Henry Sweetbaum. The families used to live in the same building, one above the other. Sweetbaum had gone to Wharton College and elected to move to London, making friends with the Plessey family. An arrangement was made to sell OKI PBXs in the U.S. under the Plessey label, and Blau was appointed to set up distribution. Sherman joined him, calling on Muzak dealers, electrical contractors, and key system distributors.

After working for Plessey, Sherman was hired by Tom Kelly, President of TIE. "TIE worked out of a converted gas station on Oak Street in Greenwich, Connecticut," Sherman recalled. (Kelly said it was actually a plumbing supply house with propane tanks outside). "Gil was always wonderful to me—treated me like a kid brother. Both Kelly and Engles were pretty big drinkers. I didn't drink until I was twenty-six. When I joined in October of that year, I was given a salary of $18,000. In December they gave me a $15,000 bonus. Later, when I wanted to buy a house, they gave me another $15,000 and gave me a car. While I was happy with my new role and income, my father thought I was being underpaid. Even though he was an upholsterer, I always valued his opinion. He said, 'You're doing so well for them, they're feeling guilty about it.'"

Some time later Sherman was feeling "pissed" for having been passed over for a promotion and decided to resign. He was hired in October of 1974 by Steve Mihaylo, President of Inter-Tel, a new startup and direct competitor to TIE.

"When I met Mihaylo at a NATA convention," Sherman said, "he had a small table covered with a checkered blue tablecloth and a 'Key Lux' phone—but with twenty-five-pair cable when TIE's phones at the time required larger fifty-pair cable. Kelly and Engles were furious. Gil came over and picked up the car they had given me. I wanted to become part of a company with the potential to earn more money. Mihaylo presented that opportunity, although he had no money at the time. He offered me stock options and an eight percent commission selling Inter-Tel key systems to dealers as an independent rep. Selling $11 million in Inter-Tel's 'Key-Lux,' I made $880,000!"

Sherman spent $100,000 of his earnings to purchase eight percent of Inter-Tel stock before it went public. Within a few years, his eight percent was worth millions. Problems began to develop with the introduction of Inter-Tel's new SPK system. Mihaylo brought in Richard Long, former CEO of Fisk Electric's communication division and President of NATA, the industry's trade organization. Long was hired as Inter-Tel's CEO, Sherman, again, felt slighted and began to trust Mihaylo less. Before he became an Inter-Tel employee, he'd agreed to work on straight commission as an independent rep for the first two years with no "draw" or advances against commissions. Sherman felt he had contributed the most in building out Inter-Tel's dealer network and the revenue that flowed from it and deserved the CEO title.

The relationship between Inter-Tel and its principal Korean supplier had become strained, and Mihaylo began to seek out a second source. During one of their scouting trips to Korea, Sherman met J.Y. Lee, President of GST a division within Korean giant, LG (Lucky Goldstar) Electronics, and son-in-law of LG's Chairman, Kwang-Mo Koo. Lee took an immediate liking to Sherman; at one point Lee declared he was going to "marry Sherman for the rest of his life"—implying a life-long financial relationship.

Sherman decided to keep his new relationship with Lee quiet from Mihaylo for the time being. As Sherman painfully recalls: "I had valued my relationship with Mihaylo, who I found to be "very talented, brilliant, but was all for himself. This was the worst of the worst. We had just gone public. Orders had gotten canceled [due to Inter-Tel's new problem prone SPK system]. I was so worried . . . the problems were all over the place . . . and [Mihaylo] went into all these other businesses, airplanes, retail hardware stores, etc. He was defocused.... I wanted him to leave ... stay at home and let me run the business.... I felt that I had built it. I felt it was my company and I wanted to run it."

Steven Sherman (R) and J.Y. Lee of GST sign Vodavi agreement.

During the time of the SPK problems, a rumor circulated in the

company that Mihaylo, from his phone, would casually "eavesdrop on
employees' phone conversations." No one interviewed could confirm
the rumor, but it added an extra element of unnecessary strain.

With financial help from J.Y. Lee and some of his own capital,
Sherman decided it was time to start his own company. He incorporated Vodavi Communications Systems, Inc. in 1983, and brought
Gehardt Klaiber along with him from Inter-Tel. Klaiber previously
was Division Vice President and General Manager for Siemens
before joining Inter-Tel as Senior VP. Sherman had known Klaiber
while at TIE and helped recruit him to Inter-Tel. Sherman also
brought on Kent Burgess to help start the new company. Burgess
would become Senior Vice President and Secretary. He'd been
working for General Telephone of the Southeast since 1973, selling
PBXs and key systems to the GTE customer base. He was surprised
to be named "Salesman of the Year" and was quickly promoted in
1975. That same year, he was relocated to GTE's headquarters in
Durham, North Carolina, and made responsible for all Federal and
Civil government accounts in the Southeast. In 1979, he was
recruited by United Telephone in Florida where he was product
manager responsible for selection and implementation of all PBX
key systems and data products for the company. At a UnitedTel
conference in Florida, he ran into a friend, Mike Stewart, formerly
National Sales Manager for GTE's Automatic Electric. Stewart, at
the time, was a rep for small telecom products he was presenting to
UnitedTel. In a side conversation, Stewart discussed a possible
opportunity with a new company, Inter-Tel, and recommended
Burgess to Sherman.

Burgess was invited to fly out and meet with Sherman who
introduced him to Mihaylo. As Burgess recalled: "After the interview,
Sherman said, 'We'll get back to you in a week.' I didn't hear
anything for two weeks. I saw the regulated telephone as a dying

industry and thought maybe it was time to make a change. However, when I received an offer from Sherman, I told him, 'I am not sure I want to work for you. You said I'd hear from you within a week and didn't get back to me.' He laughed and said I was the first person to refuse him. After some negotiation, I accepted their offer of Product Manager for the SPK. This was during the time of Inter-Tel's SPK problems. I left in late 1982."

Vodavi opened an office in nearby Scottsdale, Arizona, and within months other key players at Inter-Tel—Chuck Kelly, and Larry Steinmetz—also jumped ship to join Sherman's team. Steve Mihaylo reacted to the proselytizing of Inter-Tel's valued employees and filed suit against Sherman. The suit lasted for years and was eventually dropped but the relationship between the two individuals was irreparably damaged. To Mihaylo, Sherman's actions were a breach of trust and loyalty. To Sherman, "It was just business, and I was surprised at Mihaylo's reaction." Vodavi's digitally-based hybrid key systems would target the same market segment as Inter-Tel's, but Sherman had a different strategy.

Sherman would also hire Michael Stewart, who had been working out of Tampa and was the account manager responsible for the airlines and airport before becoming National Sales manager for GTE's Automatic Electric. When they first met, Sherman was an independent rep for Inter-Tel. GTE hired Sherman to teach Stewart how to become a sales rep. Stewart was tall, good looking, raised on golf, and, as Sherman describes, a boardroom salesman who could get appointments with anyone—and did.

Sherman later added, "I was just a forty-year-old kid, worth about $10 million at the time, and Stewart and I went to North Supply and walked out with a $60 million exclusive contract for a 'private label' hybrid phone system called 'Premier' to be made by Vodavi."

Before the Interconnect market arrived, North Supply and Graybar were the two largest supply houses providing mostly installation materials, cable, racks, and single-line telephones to the regulated telephone industry. Once the market was open to competition, both companies expanded their product lines by offering telephone systems.

Armed with a $60 million contract, his own funds, and an investment by L.G. Electronics through his friend, J.Y. Lee, Sherman took Vodavi public in November of 1983, just ten months later. At the time, Vodavi had only about a dozen employees. Bear Stearns and Rooney Pace, a smaller boutique investment firm, conducted the underwriting. The investment banker, Sanford Bernstein, also participated in the deal, taking about half the offering. The public offering raised $15 million for Vodavi, one of the larger public offerings for startup private telephone manufacturers and importers in the early '80s. In 1987, Rooney Pace agreed to leave the securities industry after an investigation by the Securities Exchange Commission found numerous violations in three of its public stock offerings. Vodavi was not named as one of the offerings investigated. The analyst who brought Sanford Bernstein into the deal was Steve Churst who first recognized the investment potential of MCI and had attended the same college in New York as Sherman. Churst would later become Vodavi's VP of corporate development and would lead Vodavi into several acquisitions.

Klaiber remembered an effective Siemens sales manager, Glenn Fitchet, who had worked for him and been promoted to National Sales Manager in 1978. Fitchet left Siemens to join a startup company, Valcom, in 1981. In early 1984, Klaiber hired Fitchet as Vice President of Sales for Vodavi.

Rather than establishing a host of independent dealers as he did at both TIE and Inter-Tel, Sherman targeted larger accounts. After securing the North Supply contract, Sherman approached Centel, one of the largest independent telephone companies, with a similar concept of providing them with a private label system. A "Take or Pay" contract was negotiated that would continue to fuel Vodavi's revenues. The first year ending 1983, revenue was a modest $756,000, and they reported a $1 million loss. By the end of the second fiscal year, 1985, revenue had grown to $71.3 million with net profit of $2.5 million. Sherman next sought a joint development with PacTel Products, a subsidiary of Pacific Telesis, one of the newly formed "Baby Bell" companies resulting from the AT&T divestiture. In the agreement, each company would own fifty percent. Vodavi would supply private-labeled single-line phones, designer phones, and rugged handsets with protected cords for pay telephones to the new joint venture.

Vodavi's product line would grow to include Triad which offered up 384 ports (stations and trunks), Starplus, Infinite, and IVR systems (Interactive Voice Response). Vodavi would also become a major supplier to Executone, which, became Vodavi's largest customer. The following year ending December 31, 1986, Vodavi revenue would dip eighteen percent to 58.4 million with a reported loss of $5.7 million, and the following year reported losses were $7.9 million due to "write downs of acquisitions." Sherman should have had enough on his plate to satisfy his drive in building a formidable company. But no, he had other objectives. In November of 1987, Vodavi announced signing a $45 million contract with Siemens Information Systems (SIS), adding Vodavi systems to complement Siemens's line of PBXs. However, when the Siemens Corp. officers found that the U.S. division was losing money, they canceled the Vodavi contract with their U.S. subsidiary.

Contel, one of the largest independent telephone companies in the U.S. at the time, had purchased Executone during the early '80s

and had not been happy with the operating results. Executone sales
had declined by a third to $200 million a year with reported losses of
$26.3 million in 1987, $15 million in 1986, and $29 million in
1985. On October 19, 1987, the market crashed, dropping some 500
points (today, 500 points might be considered a mere "correction").
Many corporations began to look for ways to trim costs and sell off
money-losing divisions. At Vodavi, concerned with the financial
stability of one of its largest customers, Sherman sensed another
opportunity and thought that Contel might be looking for a way to
dump Executone. Sherman and Churst called on Contel and they
were right. Contel was eager to dispose of the asset acquired only a
few years earlier, and had put the Executone division up for sale.
Three firms were potential bidders: TIE, ISOETEC, and Vodavi.
TIE had reached an agreement with Contel, but the deal fell apart
when the price of TIE's stock dropped and they scrambled for
funding. Sherman and Churst met with Contel's President, John
Lemasters. The three of them reached an agreement that Vodavi,
whose revenues at the time were about $70 million but losing
money, would buy Executone for $60 million. As Sherman recalls, "I
was really surprised. They displayed an attitude of, 'Just take it.' No
cash was involved."

The other major supplier to Executone was Kessman's ISOETEC,
a new player in the field. Kessman had learned that Vodavi was
intending to purchase Executone from Contel and didn't want to be
left out in the cold. With Vodavi's acquisition, Executone would
likely drop the competing ISOETEC line, or only limit dealer
purchases of their products to satisfy moves and changes in servicing
the ISOETEC customer base. Stan Blau, at the time Vodavi's CEO
and long-time friend of both Kessman and Sherman, met Kessman at
a NATA convention. The two began discussing the pending Executone
purchase. Kessman suggested a three-way deal with both companies

jointly buying Executone. Sherman really didn't want to manage a large company. He preferred to build companies. In addition, a merger of Vodavi and ISOETEC would spread the risk and free Sherman to do other deals. In December of 1987, the announcement was made: The two companies would merge and buy Executone, changing the name to Executone Information Systems.

It took the better part of a year from the date Churst and Sherman reached an agreement with Contel for the three-way merger to be completed and approved by shareholders in July of 1988. Kessman was quoted in the October 11, 1988, issue of *The Arizona Republic*, stating, "The merger was one of those things where you wake up one night and say, 'Hey, we should do this!'" The article continued: "He knew Vodavi was talking to Contel, and he knew Steve Sherman well and saw a synergy. Kessman, 41, at the time played a key role in blending the two companies that were losing money [into] one profitable firm to make money—making a telecom giant." Because Vodavi had chosen to sell its products through third parties, Kessman claimed, "Vodavi had a good product but lacked cohesiveness and did not have direct influence over sales.... They were not in control of their destiny." But Kessman was a bit optimistic, stating in the same article, "running a business with $362 million in sales and reporting a loss of $40.6 million in 1987 for the combined companies and that the company should report a profit for the first quarter in 1989."

On the surface, it made sense. One analyst thought the reduction of Executone's heavy overhead and combining of sources would likely result in more profits. It was easier to forecast than perform the execution. The two founders could not have been more dissimilar. Some close to both said, to put it mildly, they did not enjoy each other's company. Kessman came from a financial background—a numbers man. Sherman was a deal maker, relying on personal

relationships, and considered the Vodavi products made by LG in
Korea of better quality than ISOETEC's, made in the Dominican
Republic. Kessman would become President and CEO; Sherman
would become Chairman of the Board. Blau, who had become the
president of Vodavi, was given a well-paid position on Executone
Info. System's Board for his efforts. Sherman would be paid a
"severance fee" of $2 million to "transition out." In 1988, Sherman
would resign Vodavi to "Pursue the Creation of New Entities." As
Sherman states, "Kessman lacked vision, employing former ISOETEC
key people in positions of responsibility over those from Vodavi.
They gradually phased out Vodavi products in favor of the ISOETEC
products which had more advanced features. Even my father asked
me, 'Why did you do that?'"

It took until the end of 1991 before Executone achieves a small
profit of $1 million on sales of $243.6 million. The company would
continue to struggle to achieve revenues above $300 million with net
profits ranging between 1.3% to 2.5% of sales until suffering a
significant loss of $37 million in 1995. The loss was attributed to
"discontinued operations" from the sale of Vodavi which Sherman
repurchased. He simultaneously resigned from Executone's Board.
Sherman had bought back Vodavi for $10.8 million and, again, took
the company public, resigning in 1998. Between the time of merging
Vodavi with ISOETEC and Executone and its repurchase, Sherman
had formed Sherman Capital Group, a merchant Banking organization.
Sherman Capital made investment in and founded several other
companies including: Main Street and Main (the largest franchisee
of T.G.I. Fridays), Purple Wave, Telit Communications, AirLink
Communications, Novatel Wireless, Ekso Bionics and Imtrik, a
Montreal based company.

CHAPTER 32

ISOETEC: In Search of Excellence Through Employee Commitment

Alan Kessman's entry into the telecommunication world happened when he joined TeleResources (T/R) in Greenwich, Connecticut, as their comptroller in 1973. Kessman rose to become the VP and CFO. He resigned when the president failed to honor his agreement. Kessman was hired to become President of ROLM Midwest in Chicago. He was asked for his opinion of his former employer by the investment firm Hambrecht and Quist. Bill Hambrecht was appreciative of the input and invited Kessman to call him if he ever decided to do something on his own.

Jarvis Corp. had sold their PBX business to ROLM, and Mike Jarvis became President of ROLM Atlantic. Kessman and Mike Jarvis became close friends. With the passing of his brother, Carroll, in 1981, Mike resigned from ROLM and assumed the role of Chairman of the Board for the remaining Jarvis Corp. that sold and leased smaller systems.

After IBM agreed to purchase part of ROLM, Kessman and Mike Jarvis concluded neither one fit the IBM mold. Jarvis suggested they form a company to manufacture and distribute an advanced featured hybrid Key system targeted at the 100 lines and under market. The two former CEOs decided to form ISOETEC in

1983, and recruited Shlomo Shur from TIE to head up software development. The acronym stood for "In Search Of Excellence Through Employee Commitment." Their proposed system would contain many of the features that ROLM had pioneered. ROLM had introduced their VSCBX in the late '70s to address the fifty- to 144-lines market. Due to its minicomputer architecture, it was still considerably more expensive than other products addressing the same market. Only the SPK product by Inter-Tel had targeted the same market with a hybrid system employing similar ROLM-like features, but their system was problem-prone. It would take two more years before Inter-Tel would introduce the GX-120 that reliably provided feature-rich capabilities replacing the SPK. There was a window of opportunity, and Kessman remembered Bill Hambrecht's invitation to contact him when contemplating a new venture. He placed a call, and Kessman and his team founded their new company in 1983 with $5 million in venture capital funding from Hambrecht & Quist.

ISOETEC decided to have their product made in the Dominican Republic and began distributing through independent dealers and Executone. Two years after the company's incorporation, they purchased Jarvis Corp. In four short years, Kessman grew ISOETEC to $101 million in revenue, reporting profits of $1.5 million in 1986 and $6 million in 1987. The company issued its first public offering in October of 1987. As Kessman remembers, "ISOETEC was the next to last company to go public as the stock market crashed on October 22, 1987." Two months later, ISOETEC would agree to merge with Vodavi and the two would purchase Executone. The agreement would be finalized six months later with Kessman assuming the position of President and CEO. ISOETEC had a short, dramatic history in the industry, but its founder and CEO had a long one, assuming major roles in TeleResources, ROLM, Executone and ISOETEC.

BOOK SIX

Israeli Manufacturers

CHAPTER 33

Telrad

Several ROLM distributors would occasionally meet to share success stories and common concerns. Those of us who had a thriving business selling smaller key systems had a common complaint. More dealers were cropping up in our areas basically selling the same product, sometimes with only a small cosmetic change. All of us had some exclusivity we enjoyed with the PBX manufacturers, but that was not the practice with our key system suppliers. Suppliers like TIE, Toshiba, Panasonic, Iwatsu and industry supply houses of North Supply and Graybar began to sell to most any Interconnect or electrical contractor. A few of us began to explore having our own product for the under 100-line market. Stan Blau introduced us to a small Israeli company, Telrad, which was owned by the largest conglomerate in Israel, Koor Industries. Telrad was operated through their trading division, Solcor. We liked Telrad and the direction they were taking with a product they had in development based upon micro-controllers.

Four of the ROLM distributors—Coradian, Jarvis Corp., Electronic Engineering, and TSI—along with Blau, formed Pentacom, Inc. In early 1978, Marty Liebowitz, the President of Electronic Engineering, and I met Mike Shalit, our contact at Telrad, and worked with their engineering team at their Israeli facility on feature sets and operational capabilities. Each of the four of us would have exclusive distribution in our respective areas, and Blau would establish exclusive agreements with other ROLM distributors and independent distributors in other cities. We brought in Bill Jacobson from TeleResources to help with the design of the system. We called it the KeyBX, as it had some PBX features and incorporated an LED display in each of the telephones. It was offered in configurations of eight by sixteen, twelve by twenty-four, and sixteen by thirty-two (trunks and stations). The Israeli engineers were always accommodating, constantly seeking our input during its development. Working on our own product with Telrad was exciting and fun. Shipments began in mid-1979. TSI took the first twenty shipments. The other Pentacom partners were smarter and waited to see how the TSI's first installations went. There were problems—both hardware and software problems. This at a time when TSI was starting to have its own cash flow issues. Telrad engineers were sent out to help correct software problems, but the power supply, also made by Solcor (now Koor Industries), continued to overheat. The other Pentacom partners delayed taking shipments or took none at all. Blau was initially successful in establishing RadioShack as a national distributor, obtaining a $20 million contract. However, due to quality issues, the contract was dropped.

Sales of the KeyBX limped along until late 1985, when Dr. Ben Eisner, formerly with another Koor company, Tadiran, became President and incorporated the U.S. operations as Telrad U.S. By then, the system was becoming dated. Eisner started in a different direction. In 1986, a new generation KeyBX was introduced, with a slick design including an interactive LCD display. To lower production cost, he outsourced a new power supply, designed in Connecticut, and introduced for the first time a low-cost station. The new power supply solved the problems and reduced its cost twenty to fifty percent from the less reliable earlier version.

Later, Telrad would introduce a new, larger system supporting 128 lines with possible expansion. The system included a new LCD display and would allow connection of single-line telephones, like those behind a PBX, which lowered the per-station cost. However, it was coming to the market at a time when the under-100-line telephone system suppliers had become proliferate. As Eisner remembers, "It was a time when the big guys, like Nortel, began putting the squeeze on its distributors to get rid of competing products. Nortel introduced the Symposium which incorporated one of the first ACD (Automatic Call Distribution) feature in a key system, and it took off like a rocket ship. We reached a 1.5% to 2% market share yet, even with the new system, we were not competitive at the lower end (smaller-sized systems). Even though we increased the number of distributors in the U.S. to between 150 to 200, we achieved less than $40 million a year in sales." Eisner would resign in 1990.

CHAPTER 34

Tadiran

Once again, Stan Blau, made the contact. In the early 1980s, I was in search of an offshore company that either had a telephone product or had one in development and needed U.S. distribution. I made reservations to attend the International Telephone Conference held every four years in Geneva with the intent of following Steve Mihaylo's strategy with Inter-Tel. The conference in Geneva is the largest telecommunication conference in the world. Companies spend millions on booths, some as high as three stories, to display their products and services. People have died constructing some of these booths. Before my flight, Blau had suggested meeting executives from Tadiran who were also attending. Tadiran was a subsidiary of Koor Industries. A meeting was set early at the conference. Tadiran had been distributing an earlier small PBX called the TADEX. It was an older mechanical technology and lacked sophisticated features of most PBXs in the early '80s. Its major benefits were that it was inexpensive and, supporting under fifty stations, required no cabinet as the switch was contained in the console. It was ideal for small motels, resorts, schools, and small factories. As for reliability, customers who had them didn't want to remove them. Their gem, still in development, was a very modular, digitally-based, processor-controlled, 1,000-plus-line PBX. The surprise was that it had an inexpensive KSU (key service unit), costing under

$400 and used the same telephone from ten stations to 1,000. Its modular architecture allowed the system to grow inexpensively as the user grew. Few systems in the early '80s grew without a significant hardware and software change.

I was enthusiastic about the system, which would be called "The Coral." Unfortunately, Tadiran was unwilling to relinquish control of distribution in the U.S. to an outside party. We negotiated for three days and agreed to meet in Israel after the convention. There, it was another five days of negotiations with Yossi Feldman, Haim Rosen, and Abe Atar. The Israelis were tough negotiators. We finally came to an agreement and entered into a consulting agreement for one year, during which I would manage their existing U.S. sales force that was selling the TADEX and setting up distribution for their forthcoming Coral PBX.

As soon as I arrived back home, I was asked to report to Tadiran's facility in Tampa which they had recently purchased from GTE. There, I met Pinchas Just, who would become a life-long friend. The dealer network was very weak, and the regional sales force was small and not very motivated. Two of the salespeople were turning in false expense reports—not really a surprise. One salesman confessed that the sales calls on his reports were never made. Several years earlier, while making sales calls in his car, he saw a child crossing an intersection with a real pistol in his hand. He stopped to talk the boy into handing him the gun and was shot in the stomach. Since that incident, he had been frightened to make any sales calls and had been fabricating his reports. One sales hopeful enthusiastic and loved working for Tadiran. He had recently given his wife a new Cadillac. This was 1984, and Cadillacs were still a sign of success. He, however, was not. His expense reports included mileage reimbursement—lots of miles driven—plus receipts for car rentals. That was a red flag. With little investigation, it turned out that the cities he visited on his call reports were consistent with his son's high school baseball schedule to other cities—mostly cities too small for a distributor's survival. We hired a VP of Marketing and a VP of sales, and changed out the sales force.

It became a challenge to induce new distributors to take on the older TADEX PBX, as it just wasn't sexy with features, and it required "trunks" used on most PBXs rather than "business lines" from the central office. Typically, "trunks" cost more per month than "business lines," depending upon the operating company. Regardless, those customers who bought them loved them. The TADEX was so trouble free that they didn't want to change it out as newer, more feature-rich systems became available.

Bell South became a distributor for both TADEX and Coral Tadiran systems, placing hundreds of TADEX systems in the Broward County School System. Bell South would wind up selling 250 of Coral systems. Bell South also sold Dade County a large Coral system, only to have the contract overturned when Avaya contested the contract. By the time the Coral was introduced to new distributors in 1985, the market had matured, and most major PBX manufacturers such as Nortel, ROLM, Siemens, and NEC had locked in major distributors who would risk losing their exclusivity if they picked up a competitive line. The selections of qualified distributors had shrunk while financial leaders were forecasting a reduction in revenues for the telecom industry.

Some independent distributors such ICS in Texas, started in 1983 by Dannie Simons and his wife, did quite well with the Coral. Even though they started their company more than ten years after the industry began, ICS also encountered problems with an AT&T operating company—Southwestern Bell Telephone (SWBT). Simon recalls, "We always had problems with the protective couplers provided by SWBT. The devices often ran 'hot' and wound up dropping calls. Of course, we'd call SWBT, and when they finally called back, they'd claim there was no problem with the protective couplers—which was BS. Our guys wound up taking them apart to locate the problem. They [SWBT] also got nasty installing the things. With one customer in Waco, we had an installation going in at Nash Chevrolet. SWBT decided to install the connecting devices outside, under the eave of the roof at the highest point in the building. When we requested they install them by the main frame [distribution panel in the equipment room] next to the ICS system, their reply was, 'Our rules say we can put them anywhere, and that's the place we chose!' A bucket truck was needed to put them there and to remove them. After the installation, the customer said to me he'd never do business with the phone company again."

During an installation at another customer site, Simons overheard a second level SWBT supervisor tell the Bell technicians to cut the cabling flush to the conduit, rendering both the cable and the customer's conduit unusable. Simons continues, "Toward the end of one installation, as Bell was removing their equipment, a Bell technician walked over to the operator and with his pliers cut the headset cord she was talking on at the time."

Simons recalled selling approximately 300 Tadiran systems in Texas. One of the initial sales contained a 120-party conference bridge that Tadiran helped engineer. As the Texas market covered a lot of territory, Simons bought a thirty-eight-foot Itasca motorhome and rigged it with telephone equipment, including a Tadiran Coral SL, which was used for live demonstrations. ICS had one competitive sale that was made to Westinghouse that included several branch locations. All the major players that were competing for the sale—Siemens, NEC, Nortel, Mitel, etc.—were unable to light a message-waiting lamp on branch office phones from Westinghouse's central voice message system in Georgetown, Texas. Tadiran was able to write a protocol to work with QSIG, a networking standard, and demonstrated the message lamp feature working in a Westinghouse branch location. ICS obtained the contract for twelve systems. They had installed Tadiran systems in fifty Texas State Bank branch locations when the company put the remaining forty branch phone systems out to bid. As led by the Bank's consultant for the project, Siemens appeared to have the bid in hand. The Siemens' salespeople and engineers had walked into the customer's conference room with their briefcases crammed with their presentation and backup data. Simons parked the ICS motorhome in the Bank's parking lot early that morning. When it was ICS's turn, Simons invited the Bank's executive outside to the motorhome for a working demonstration. Nothing beats a live demonstration. ICS won over the consultant, and the bank and got the contract.

Even at its peak year, in 2000, Tadiran U.S. would achieve less than $40 million in sales, or under two percent market share. Internationally, the Coral did well in countries like Ireland, Russia, and India, which has eighty very large Coral systems installed.

Pinchas Just recalls, "As the Coral was becoming a less effective product in the marketplace, Tadiran lacked the necessary financial resources to handle growth, research, and development of a replacement system. Senior management, Haim Rosen and Yossi Feldman, went to the Koor Board to make the request for more funds. The request was turned down. Instead, Koor sold the company to the Broffman family, brothers Charles and Edgar. They put the son of a best friend in charge of Tadiran, who closed down Tadiran's R&D efforts and tried to sell parts of the company, eventually ruining the company." Tadiran was finally merged with ECI, a large Israeli company focused on central office, VoIP, and internet products. ECI decided they wanted little to do with Tadiran and sold the company to an Israeli private investor, who sold it again to a car leasing company. Currently, employees have been reduced to forty people, two of which are in the U.S.

• • •

BOOK SEVEN

Conclusions

Telecommunication Services, Inc. (TSI)/First Communications— Part II: One Domino, Two, Then Another

n February of 1978, I was planning a trip to visit TSI's recently acquired First Communications' offices in Florida. We had consolidated offices, improved some locations, and hired additional salespeople. "Why don't we make a small vacation of it and take the kids?" asked my wife. Sounded like a great idea. We'd fly in, I'd work the week and join the family on the weekend. The family was settled into a hotel near Disney World.

Arriving at the hotel after work toward the end of the week, Ken Oshman, ROLM's president, called. He must have phoned the St. Louis office to find out where I was staying. Oshman began: "We've decided to open a ROLM direct office in Chicago." I was floored— not good news. He continued, "There was never a distributor agreement executed between ROLM and TSI for a distributorship in Chicago—you expanded from St. Louis. We will need to cancel your St. Louis distributorship. However, we would be willing to issue new distributorships for Missouri and Kansas, and confirm First's distributor agreement for Florida with the understanding that you are willing to transfer TSI's ROLM Chicago customer base over to us along with any contracts in progress."

"Let's meet and discuss this," I said. "I'll fly out today."

"We've already started operations there," came his quick response. "An office has been leased, and Alan Kessman has been installed as President of ROLM of Illinois."

"Can you hold off making any announcements until I can get to Chicago and inform our employees?"

"I don't think that's possible," he continued. "Kessman is starting hiring interviews on Monday. Bob Maxfield will work with you on details of the transition. Goodbye."

"Goodbye" was right. It was the last time I would have any conversation with Oshman. At least he didn't ask for my wife and kids. His comments were well prepared—but that was consistent with the way Oshman worked: calm, concise, choosing words carefully, and probably consulting in depth with ROLM attorneys. I assumed he was right on the distributor agreement, as I could not recall amending it to include Illinois. It was an oversight by both parties.

Kessman had last been the CEO for TeleResources. I knew and had respected him while he was working with T/R, one of TSI's major suppliers. When ROLM hired Kessman, it was certain that both parties knew what his eventual role would be: President of the first ROCOs (ROLM Operating Companies), responsible for Illinois and, later, the Midwest. While TSI was investing thousands in promoting ROLM at our teleconference in Chicago, ROLM was in the wings, waiting to take advantage of the growing opportunity. TSI had made a considerable investment in ROLM for Chicago: We had paid ROLM's tuition costs for sending our technicians to their technical training school, plus their hourly rate, travel expenses, etc. TSI also invested in the hiring and training of coders to use ROLM's "auto-quote" system to configure systems. Additionally, we had a

major investment in inventory and CBX spare kits that the service
and installation personnel were required to carry in their vehicles.

There had been no previous discussion of ROLM's intent to
start direct sales and service operations in any area TSI was selling,
installing, and servicing their CBXs—no written impending threat—
nor did ROLM provide some future date when they intended to
commence direct sales. Certainly, they would have liked us to sell
more PBXs and drop the PBX products like TeleResources and
Mitel that did well in the under 120-line (telephone stations)
market. ROLM wasn't cost effective in the small business market,
but that didn't concern them—which became clear in our meeting
with them at the end of our Telecom Conference we hosted in
Chicago. In fact, to my knowledge, prior to their entry into the
Illinois market, they had not informed *any* distributor of their intent
to open a direct sales office in that distributor's location. A few years
later, ROLM concluded that their New York distributor, Coradian,
which was publicly traded, was doing fine in upstate New York but
wasn't selling enough CBXs in New York City. Unable to reach an
agreement to purchase the distributor, ROLM opened a direct sales
office there. ROLM agreed to limit their first-year sales to only
"national accounts" and allowed Coradian to continue the distribution agreement for another year. After that, the distributor agreement
was not renewed.

Such was not our case. Nor was there ever a comment or
suggestion of ROLM's buying TSI's ROLM customer base in Chicago,
buying our backlog of ROLM sales or CBX work in process, buying
our inventory, or reimbursing us for the cost of training technicians—
most of which made a bee-line over to Kessman to be interviewed
for a position, probably some while on our payroll. Nor was any offer
made to compensate us for the telecom conference we recently
conducted that showcased the ROLM CBX. It wouldn't have cost

them much to work out an agreeable solution which would have resulted in less disruption to the customers and our employees. We couldn't believe that ROLM assumed TSI would willingly give up and transfer our sales backlog and CBX installations in process to ROLM without adequate compensation. Fellow ROLM distributors couldn't believe it either.

Much earlier, the distributor in Denver went out of business. Boston's ROLM distributor was soon to follow. Bob Fabricatori, the founder of Independent Business Telephones, called to inform me and other ROLM distributors that he had told Oshman they didn't have the funds to continue, and asked them to come take them over. In return, Fabricatori was provided shares of ROLM stock. The deal was done on a handshake. It was learned later that between the handshake and the signing of documents, ROLM's shares split, and the price of their stock had rapidly accelerated. The deal was based not on a dollar amount, but a quantity of shares. ROLM honored the agreement to Fabricatori's benefit. In each case, they did what they needed to continue the sale and service of their products without consulting with their former distributor.

As for TSI, we had a profitable ongoing operation and had been paying our bills on time. Our experience wasn't the same. We had about thirty CBXs in the Chicago base, and some were still having problems due to overheating or occasional technical problems. On the flight back from Florida, I reflected on what might have precipitated ROLM's actions and what the alternatives were. Was it TSI's acquisition of First Communication in Florida without consulting with ROLM in advance? Was it the comment Knox had made to the ROLM managers or his reluctance to address some ROLM customers that were having problems (for which he had been removed from his position a few weeks earlier)? Were TSI's sales in the Chicago area that far behind ROLM's expectations? Was it the opportunity demonstrated by potential customers who voiced their desire to contract directly with ROLM at TSI's recent Chicago telecom conference? Was it me? Where was the correspondence to support any of these reasons? Perhaps ROLM thought of me as an uncontrollable maverick they just didn't want to deal with. Based upon conversations with fellow ROLM distributors, TSI had to be one of their largest distributors. I took it personally.

TSI chose not to take legal action against ROLM as we didn't have a distributor agreement to sell ROLM in Illinois, and we did not want to jeopardize our distributor agreements with ROLM for Missouri, Kansas, and Florida. However, they next reduced our credit line, arguing that without our ROLM sales in Chicago, we didn't require as much. We would need to quickly restructure our Chicago operation.

We did not intend to give up our customers or abandon our backlog. ROLM refused to ship us two CBXs scheduled to be installed in Illinois State Agency offices. They assumed they would be taking over the account and would complete the installation (and collect the profit). Rather than go to battle over the customer, we went to the customers and proposed changing the order from ROLM to an Automatic Electric PBX. The State went along with the change order as their focus was more on the total operating expense than the additional features the CBX offered. ROLM was furious. Some of our Chicago ROLM customers called us confused after being contacted by them. They had warranties on their systems, and TSI had an obligation to provide them service and parts replacement.

In a phone call, Kessman assured me that he would not directly go after our ROLM customers nor solicit our sales or installation people, but added that he intended to interview and hire qualified TSI employees who came to him for employment. I couldn't blame any employee that had been trained on the CBX to explore opportunities at ROLM. ROLM of Illinois intended to subcontract out the wiring needed for new installations. Their technicians, dressed in suits and paid a salary, would provide testing and "cut-over." It was assumed in this manner they would be able to bypass the need to have union employees. TSI, with its trained union technicians, was never asked to bid on any of ROLM's installations. ROLM was growing rapidly, its stock trading at thirty times earnings, so wearing a coat and tie could be considered more prestigious to the techs—and the CBX was the best product on the market. When any issue between ROLM of Illinois and TSI arose, Kessman was quick to call and discuss the problem with me. We lost about half our Chicago employees. Although ROLM began to service some of our customers, TSI's ROLM customer base and inventory would not be transferred to them until 1981. During the next few years, they appeared a little more gracious in opening direct offices. In attempting to purchase a distributor, ROLM rarely offered more than book value—and in ROLM stock. When a sale could not be completed, they provided advance notice that the distributor's agreement would not be renewed at the end of the year.

Compath National, formerly Scott-Buttner Electric and owned by Ted Buttner, was a ROLM distributor that covered several cities in California. During '83 and '84, sales were running about $12 million. Compath had a positive relationship with ROLM whose headquarters were less than an hour away. Pat Howard, a former sales manager with Arcata, had become Vice President of Compath and had spent time with ROLM engineers defining the attendant functions the CBX would require during its development. ROLM wanted a direct operation in California and respected Howard's sales team and sales approach with customers. Being able to take a potential customer over to ROLM's sprawling campus in Sunnyvale often resulted in a sale. Oshman approached Buttner with an offer.

Whatever it was, Buttner thought the company was worth more and refused. ROLM didn't like to negotiate and elected to not return with another offer. When Oshman told Howard their offer was rejected, Howard and another officer of the company drove to over to ROLM and met with Bob Maxfield. Apparently, ROLM offered Compath $3 million to cancel their distributor agreement, and Howard returned to Compath with the offer in hand. One can assume the offer which was accepted included title to the established ROLM customer base. Buttner later sold the remaining portion of the company to Alltel, formerly Allied Telephone Company, for about $8 million. It's likely that the two purchases of Compath's assets were greater than ROLM's initial offer. TSI did not have the benefit of such an exchange.

We absorbed the Chicago reorganization costs during the year, and Van Cleave suggested we write down our inventory as some replacement parts and trade-ins had been eclipsed by increasing technology. We finished the year with a modest seventeen percent increase in sales but took our first loss since the inception of the company. By the end of the first quarter of 1979, the company was back on track and finished the year with $14.3 million in sales and $824,383 in profit. One cost item had more than doubled over the prior year. Interest expense was almost $250,000. In 1977, it wasn't even a line item.

Photo courtesy of St. Louis Post Dispatch, Feb. 22, 1979

Author demonstrating digital ROLM phone with display.
In the article I was claiming Southwestern Bell's recent rate increase
on residential services were subsidizing business phone services.

Van Cleave was convinced we needed to leverage the company to grow. In the fall of 1979, he established a $2 million line of credit with our banks, Commerce Bank and Mercantile Bank, at prime plus 1.25%. Fearful of using our credit line to fund leases rather than fuel working capital, one stipulation of the loan agreement was that we had to spin off Leasetel as a stand-alone company. At least it now felt like we had the foundation to further expand the business. We complied. Van Cleave also decided the company should enjoy some "perks" and bought memberships in a tennis club and a "members-only" restaurant. Season tickets for both the St. Louis baseball and football seasons were also purchased. These were treats I would never have considered. Van Cleave made the point that they were to be used for potential customers, hiring of employees, and entertaining suppliers. I was also enjoying the perks and embraced the idea.

The company needed to find a replacement for ROLM in Chicago, and Van Cleave was assigned the task. The company continued to sell the TeleResources TR/32 and OKI PBXs, but OKI informed us that one of their systems we had been selling would not be expandable as originally advertised. This caused us problems with two Texas customers whose OKI systems were at capacity. We also started hearing that TeleResources was having financial difficulty. We didn't think it possible, but the next year TeleResources would file for bankruptcy protection.

Returning from a week in our Chicago office, Van Cleave entered my office with a big grin. "I think we found the answer to our switch problem in Chicago," he announced. He had located a small development company in the Chicago suburbs, Eascom. Its principals had been with ITT, Control Networks, Wescom, and AT&T. Sounded good so far. Van Cleave said they had installed their first prototype PBX for a Seattle car dealer over a year ago. He'd talked to the dealer who said it was running fine. Van Cleave had some of our techs take look at it. It had a good suite of features: not as rich as ROLM but more than the TR32 with which we had been successful. The attendant position(s) employed a small CRT screen that would provide for preformatted text messages to and from the stations. It would be a first in the industry. Eascom had run out of money and was running on fumes.

We elected to acquire the company, but did not make it part of TSI. Instead, we would run it separately. We acquired Eascom for little more than assuming current trade payables, salaries, and providing the principals a royalty on future sales. Omni Communications was formed. As Van Cleave had located the company and structured the deal, he was given an option to acquire fifty percent when he was able to match TSI's investment. He developed a private placement memorandum with the intent of raising $2 million to fund the balance of development and initial contract manufacturing. This seemed to be an area of his expertise. The system was called Delta One. To minimize our investment in manufacturing equipment until the funding was completed, we decided to subcontract the manufacture of trunk, station line, and common control cards to a third party.

By the end of 1979, TSI was paying seventeen percent interest on our line of credit. Finance rates for the leasing of systems we sold also rose. We discounted our rent stream to the bank who made the credit determination on each customer. Thus, the leases were nonrecourse to TSI save for a buy-back agreement tied to TSI's depreciated wholesale cost of equipment. We would add a small amount to what the bank charged which would build over the term of the lease. Our lease proposals reflected the higher interest rates, reducing the amount of monthly savings to the customer. Making sales based upon monthly savings became more challenging. As we rolled into 1980, cash was king and everyone was playing the "float game," delaying paying creditors for as long as possible and earning interest on their cash. In April of 1980, the company was paying 21.25% in interest and our debt was increasing. The company was still making money, but our cash flow was getting tight.

We were experiencing problems with some manufacturers. Tele-Resources' Chapter 11 filing created another void in available PBXs. We had several TR32 systems in our backlog and installations in process. Our local competitors had a field day on the news about T/R's Chapter 11 status, and we spent a lot of time trying to save those accounts. We were hungry for product. So, when the Telrad Key BX began shipping from Israel, TSI took the first dozen systems while the rest of Pentacom partners waited to see how our initial installations went. Typical of any new system, there were problems. Telrad was responsive; they flew out engineers to help us. With the plant in Israel, engineers at the factory in Lot could work on software patches during our evenings and electronically ship them to us, allowing us to install new software the next morning. Still, the extra labor costs in maintaining the systems added to our operating costs. Then we began to receive the first Delta One systems from Eascom. Two were shipped to current TSI Texas customers for OKI system replacements. Two went to St. Louis customers. We decided to hold a sales conference with all sales and branch managers with the intent to rally the sales force around the Delta One PBX. The best way to describe the customer reaction and acceptance by our salespeople was demonstrated at the sales conference in St. Louis a few weeks later. The salespeople were eager to learn more about the Delta One. Steve Bell, VP for Dallas and Houston began:

"Let me list some of the new features of Delta One for you. There's: Surprise Conference" (laughter from the group) ". . . and Music-Over-Talk." (More laughter)

The St. Louis Branch Manager, Ed Dressel, chimed in: "How about: Automatic Disconnect, Phantom Caller, and Call Forwarding to Mars." (Mild hysteria)

The sales conference to promote the Delta One wasn't going over well. In fact, it was a complete disaster. Van Cleave's face was turning red. The Delta One was crap. There wasn't a sales manager or salesman in the room who was going to sell it until the problems were fixed. Van Cleave and I met after the meeting. He thought the problems could be fixed in a couple of months. I didn't share his confidence. As Steve Mihaylo said to me at the time, "You don't know if you have a $4,000 problem, a $400,000 problem, or a $4 million one." He was right. I didn't know, but concluded it was at minimum between $500,000 and $1 million—money that we didn't have.

Van Cleave had not been successful with his "$2 Million Private Placement Memorandum" to fund the development. I don't recall hearing about even a nibble of interest and was ready to write it off when the banks asked for a meeting. About $300,000 had been transferred from the TSI checking account to Omni/Eascom for deposits on "future systems." The deposits were being used as Omni working capital. The transfers didn't hurt TSI's balance sheet at the time they were made, but those deposits were now at very high risk. In this industry, no distributor made advance deposits on systems that hadn't been manufactured. Where a distributor may have a credit issue with a manufacturer or vendor, the distributor may be asked for payment on delivery or partial payment in advance of shipment. We weren't in that situation yet, but we were getting close. With a little more investigating, we learned that the one-year-old prototype Eascom switch at the Seattle car dealer had been a sweetheart arrangement. The car dealer was an early investor in Eascom and had every reason to provide a glowing recommendation. Shortly after the problems surfaced, Van Cleave submitted his resignation. "Most of what I intended to do had been accomplished," he said, "and it was time to move on." Omni floundered into bankruptcy and was taken over by Richard Scully of Courtesy Communications.

Regardless of all our problems, TSI finished fiscal year ending October 31, 1980, with slightly over $18.5 million in sales and $871,000 in profit. Cash was still king, and we were wrestling with cash flow problems and customers wanting to delay final payments on their contracts. In December of 1980, the prime rate had risen to 21.5%—the highest in U.S. history. Interest on our credit line had risen to 22.75% and was choking the company. Even though the company was still profitable, with the raise in prime rate and recent cancellation of the ROLM distributorship in Chicago, the banks became concerned and more involved—visiting our office weekly to review our bookings, receivable status, work in process, and cash position, all of which had been used as security for our loan. Our credit line agreement was coming due the end of January 1981.

After dumping Omni/Eascom, we cut headquarter staff, eliminated unnecessary perks, and began cutting payroll. Years earlier, TSI had sold a large ROLM CBX to CPI, a division of Chromalloy Industries. Allan Essman, president of CPI and a strong supporter of our company, offered to help. CPI loaned the company $300,000 with convertible options and restructured our accounting staff to focus on accounts receivable. Our cash position began to improve, but some suppliers late in the year complained of checks bouncing and placed credit restrictions on the company. This was confusing, as the company's checking account showed we had available funds. Some employees began to experience the same. We had missed a monthly interest payment. We learned later the banks were positioning themselves, deciding which checks they wanted to process while allowing our checking account balance to increase. If the banks determined we needed to pay a supplier/manufacturer to complete work in process and collect payment from the customer or lease company, they would process the check. In other cases, the banks might take more time to clear the check or return it regardless of the fact TSI had the funds to cover the check. The banks even decided to not process the employees' tax witholding checks.

Problems seemed to be coming one after another. While trying to manage our cash flows, another one arose. We discovered that our VP in Texas was stealing from the company, using several different methods including approving payment of invoices from factitious installation companies he created, inflating expense reports, and personally benefitting from collections on delinquent accounts. He was fired, but I was informed years afterward that he repeated the practice with another company. Fortunately, all TSI officers with signatory authority were bonded and the bonding company covered our losses.

Digital Telephone Systems (DTS) became aware of TSI's cash flow problems and suggested a meeting. As mentioned earlier, through the efforts of Dorsey and Knutson, DTS 1000 had become the primary PBX system for Executone and its dealers. DTS had since been acquired by Harris Corp. In the areas TSI served, the Executone dealers didn't offer much in the way of competition. Harris/DTS wanted more penetration in the same areas. It was a good fit with a couple of exceptions. The arrangement solved our current cash problem and provided a viable substitute for ROLM in Chicago and the TeleResources' TR32, now in bankruptcy. Our Texas offices were also in desperate need for a larger PBX. Don Greene, president of DTS, and some Harris executives reviewed our books and proposed an offer. It wasn't a great one—$2 million less than what Chromalloy had offered three years earlier and a large percentage on the back end, tied to performance. However, the offer was a lifesaver for the company. The DTS 1000 was not as feature rich as the ROLM CBX but offered more capability than the TR32 and had a much larger station capacity. With the acquisition by a competitor, ROLM would most likely cancel our distributorships in Missouri, Kansas, and Florida. A brief term sheet was outlined and agreed to. Harris informed TSI's banks, who were delighted to learn that an agreement for the sale of TSI had been made. They agreed to forestall collection efforts on TSI's loan. Greene was flying back to the Harris headquarters in Florida to prepare a formal purchase document. Before catching their flight, Greene said that they would like to meet with our technicians in Florida. The Harris/DTS people arranged to meet with several TSI/First technicians and installation managers at a bar after work. As the evening with several drinks progressed, the First technicians began to debate the merits of the ROLM CBX versus the DTS 1000. The discussion became heated. The following morning, I received a disappointing phone call from one of the Harris executives. Harris had concluded that converting TSI and First technicians and sales management from supporting ROLM to Harris/DTS was too much of a task. The deal was dead. Within hours, the banks made their move.

With no attempt by the banks to renegotiate the loan, a VP for Mercantile Bank, called me at home after midnight at the end of January 1981. The two banks had decided to call in the loan—now due and payable in total. They hired a security firm that had padlocked all doors at each branch during the night. A security guard was to remain at each branch twenty-four hours a day. Employees were denied access to get personal items. In a subsequent conversation with one of the bank's VPs, he told me the move was "intended to put more pressure on our suppliers to extend the company additional credit." The statement made absolutely no sense.

Just what were my options? Were there any legal remedies? And who was liable? I could fault ROLM for their precipitous actions that negatively impacted the company, but they were probably within their right to open a direct office in Chicago. How about Van Cleave's creation of a corporate staff, negotiating the line of credit with the banks, and subsequent purchase of Omni/Eascom? No, I had supported every one of his decisions, including the perks. What about the banks and their decision to not renegotiate the loan? No, the loan agreement gave them the right to call the loan—upon demand. The VP in Texas? No, his theft was insignificant in contrast to our debt. Trying to fault the Federal Reserve bank for raising interest rates was laughable. There was only one place to put the blame—me. I was at the helm and put the ship in dangerous waters. The banks had perfected their security interests in all of TSI and First Communications' assets. We filed for bankruptcy under Chapter 11. Within a few weeks of the filing, I received a call from Ed Spievak, the lead attorney for our association, NATA. NATA and several of its members, including TSI, had filed an antitrust suit against AT&T. Spievak asked, under TSI's current circumstances, if we would be willing to transfer our position in the suit to a smaller company in Springfield, Missouri. At the time, I didn't think we had anything to lose by agreeing to his request. Years later, I learned that TSI's portion of AT&T settlement in the case would have amounted to between $375,000 and $400,000—funds we certainly could have used.

Lester Miller was the president and owner Contico, a St. Louis-based company next door to TSI's headquarters on Warson Road and to whom we had sold a telephone system. My wife and I had previously purchased Miller's home in Country Life Acres, built in the early '20s by baseball legend Branch Ricky in the St. Louis suburbs. Apparently, Miller had once been in a similar situation. He provided a suggestion: "St. Louis bankers don't understand what a 'rainy day' is. Just throw your keys on the bank's table and say, 'You take it—it's all yours!' They don't want to run any company and will negotiate something with you."

He was right, but throwing the keys on the table was unnecessary as the bank already had all TSI's assets, keys included. We worked out an employment contract. I would work for the banks for a period of one year, selling TSI's and First Communications' customer bases, completing the work in process, installing the backlog, selling the inventory and all of the company's other assets for the benefit of the banks. Our remaining personal assets were pledged as security while working under the bank's contract and were to be returned upon repayment of the loan.

The bank agreed to pay for a handful of TSI employees to help liquidate the company's assets. We approached our task like a business. The sale of TSI and First's customer bases was made easy by packaging our inventory needed to service those customers. Selling the ROLM inventory in Chicago to ROLM was considerably more difficult. They didn't want to pay any more than their manufacturing cost. Finally, the bank agreed to a figure less than our cost and less than what I would have approved.

We created an informal company and called ourselves MerCom— an amalgamation of Mercantile Bank and Commerce Bank—and advertised in the local paper: "Complete Business Telephone Systems for Sale at Wholesale Prices—labor included." We added a small markup to our inventory cost and packaged the system and telephones to customer requirements. We knew which technicians were qualified to install and service the systems, and arranged for them to meet the customer, review the installation, and provide an installation and service estimate to be paid directly by the customers for their labor. Some who came in to make purchases had previously been called on by our salespeople, but had not yet signed a contract. We sold all four TR32 systems in inventory and several key telephone systems for cash before the banks required us to stop using the newspaper ad. The banks' names were not mentioned in the ad, and we were bringing in more cash for our inventory than what was on the books. Regardless of their logic, the practice was stopped after we had sold a significant portion of inventory.

About thirty percent of TSI's inventory in Chicago was lost or stolen. All of the KeyBX and TR32 inventory was missing. February in Chicago could be brutally cold. With no heat in the building, former TSI technicians learned the guard on duty liked to keep warm drinking whisky. Gifting the guard with a bottle of booze, the techs were able to remove several thousand dollars of inventory they needed to service former TSI customers. Convincing the customers in TSI's backlog to fulfill their contracts and completing work in process was more difficult. Many felt they would wind up not having to pay the balance of their contract, or thought they should be able to receive their deposit back. The company's Chapter 11 status was a major help for the banks. None of the customers wanted to take the time to petition their position before the bankruptcy court, and all backlog orders and work in process were eventually completed.

These thirteen months weren't good ones. I had lost once-thriving companies with a healthy net worth and equity in two 10,000-square-foot office buildings that housed the Chicago and St. Louis branches. Titles to the buildings were transferred over to the banks in a "quick deed" sale. Both my parents passed away during that period. Shortly after moving to St. Louis, my wife's father passed away. With two children in a new town and only a few friends, my wife had been fighting depression. With my hectic work schedule, I was totally unaware of the fact. Not knowing what my battles would likely be when the banks called in their loan, we had sold our home in Country Life Acres and invested the equity in tax-free municipal bonds at the top of the market—paying between seventeen and twenty-two percent interest. Some government-backed bonds were purchased at a discount.

Several months after working for the banks, the lead contacts from each bank paid me a joint visit. They seemed pleased with our progress of repayment at the time, and stated: "We'd like to see Leasetel's books."

"Why? Some time ago you asked that we spin off Leasetel as a precondition of our loan, which we did," I replied.

"There may be assets in Leasetel that we can use, and it will help your relationship with the banks," was their answer.

"Can you be more specific on how it would 'improve my

relationship' with the banks? Are you saying you would be willing to renegotiate the loan you called in?"

"That's not what we meant," they said, but they didn't explain further.

"I'm not clear on what benefit we get by turning over Leasetel to you. If you can provide any security interest the banks have in Leasetel, I'll be happy to comply. Until then, Leasetel's books are not available at this time."

The banks never asked again. Over the period of a year, we repaid the banks, and our remaining personal assets that they maintained as security were returned to us. The banks dragged their feet until our attorney contacted them. CPI, who tried to help the company earlier, offered me a position as sales manager to start a small Interconnect operation for them renting Iwatsu phone systems. Renting, as contrasted to outright sale or a fixed term lease, was an interesting approach that I thought would bear fruit, but I didn't have my heart in the work. Starting any new company in the close-knit St. Louis financial community didn't seem probable. We decided to move, pulled the kids out of school a month early, and bought a mini-home. We travelled the Southwest, looking for the next place to live. As the sole shareholder, maintaining control of Leasetel turned out to be a godsend. Since Leasetel had been spun off from TSI, it was ignored by TSI's financial people. No one in our finance department had been following up on its rental continuations, purchase options, and reserve—all of which provided adequate income for our family for several years. For six months we travelled the Southwest enjoying the adventure and visiting friends. No more interest payments, mortgage payments, payroll, etc. We were free spirits on a vacation, but one with a mission.

Finally, we decided to buy a home in the wooded area of Rancho Santa Fe in North San Diego County. I agreed to work for six months as sales manager for Dorsey and Healy who had purchased three Executone offices, one in San Diego. My friend Stan Blau once again offered assistance by setting up an introduction with another Israeli Telephone manufacturer, Tadiran, covered in the previous chapter.. We entered into a one-year consulting agreement in which I would assist in the introduction of a new digital phone system in the U.S. The agreement was renewed five consecutive times. What a wonderful group of people to work with.

In 1994, I was offered a consulting position with a relatively new San Diego tech company, QUALCOMM, Inc. After twenty-three years of mostly working for myself, now I was in a corporate climate and had to adjust. It was a great company with very bright people, fantastic benefits, and challenging work. Being in charge of business development and sales for one of the company's divisions couldn't have been more satisfying. Initially, while in a satellite group, my work consisted of making presentations about QUALCOMM's CDMA wireless technology to cellular operators, satellite operators, phone companies, and regulators in foreign countries. I retired happily from the company in 2002.

CHAPTER 36

Thomas F. Carter, Part II: The Fight Ends and Retirement Begins

C arter had built a thriving two-way radio business from his office on Greenville Avenue in Dallas. He had once told Wallace Hammond, Carter's son-in-law, that while the suit with AT&T was going on, he had built and installed over 5,000 Carterfones. During his fight with AT&T, Carter was forced to raise additional capital by bringing in other investors, eventually losing control of the company. His continuing battles with AT&T had reduced his once successful mobile radio and electronics firm of 100 employees down to a handful. When the FCC decision was finally made in Carter's favor, the Carterfone device had become obsolete.

After the FCC decision, Carter started TFC, a small Interconnect company working out of the same location. As Hammond states, "I went to work for Carter in early 1974.... He had about ten employees, and Helen, his wife, was the secretary. The company was over $100,000 in debt and it was not being serviced. Carter did not believe in the PBX market and only concentrated on the smaller key system sales, mostly TIE products. During the next three years we paid off the debt and were up to about thirty-five employees. But Tom was dragging cash out of the company without contributing much, so we wanted him to leave the operation and let us run it for him."

Carter didn't agree to the proposition, and the two parted company. Hammond formed his own company, Telephone Technology, Inc. (TTI), which also did installation work for others including UCS. He eventually became employed by UCS as their VP of Operations.

When Carter and his stockholders could not reach agreement on which direction to take the Carterfone company, he sold out. Shortly thereafter, the remainder of the Carterfone stock was sold to Cable & Wireless, Ltd. Hammond recalls, "As the Carterfone business ran down, he teamed with Jack Goeken who had been pushed out of MCI by Bill McGowan. They traded on their two names; the business appeared focused on selling long-distance dialing services, but I don't think they had any revenues. The business faded away in about three years." Carter had fought through a decade of agency hearings and court trials where he stood virtually alone against the giant AT&T. Other industry giants in electronics and communications had refused to back him, telling him "you can't win." Carter received about $500,000 in settling his suit early with AT&T ($450,000 from AT&T, $50,000 from General Telephone), most all of which Carter claimed went to pay his attorneys. Tired of his ten-year battle with AT&T, Tom Carter moved from Dallas to retire in Gun Barrel, Texas, stating in 1984 that the move "added ten years to my life."

"I was beat," he explained in an *Inc.* interview, "mentally, physically, and financially. You think ten years in court is fun? It sure wasn't. I had ulcers, and I was on tranquilizers."

Thomas F. Carter

GUN BARREL CITY, TEXAS

"We Shoot Straight with You," reads the town's motto. Once known as the "Old Bethel Community," the town likely acquired its name back when Clyde Barrow and Bonnie Parker hung out in the area during the 1920s and 1930s. It was considered a safe backwoods place during Prohibition. One resident, Mr. C.L. Wait, could have also contributed to the town's moniker, as he was known for sitting at the window of his house with a shotgun sticking out the window— to deter those he deemed unwelcome on the narrow main road. As Carter's veterinarian friend and neighbor, Damon Steven, stated, "The town was most likely named such as the main road through town was straight as a gun barrel." Located fifty-five miles southeast of Dallas, Gun Barrel is a small town, incorporated in the late 1960s so it could legally sell beer and wine. It sits on the edge of the Cedar Creek Lake, a reservoir completed a few years prior to Carter's arrival. By 1980, the town claimed a population of 2,118. Eventually swelling to 6,000, the town has become a second home to many in the Dallas area largely due to the lake's 220 miles of shoreline, the fourth largest in Texas.

Even if he was "beat" and on medication, Carter quickly inserted himself in the community with energy. His friends found him "affable," "an extrovert," a "great person, natural salesman," "someone you couldn't say 'No' to," as Damon Steven said. Carter was often seen in western clothes with dark-rimmed glasses and a white Stetson that he would tip toward the ladies. Several of his friends thought that Carter started the Rotary Club in Gun Barrel. If he didn't, he was certainly instrumental in recruiting members. Carter built a radio tower and shop next to his home in the town, renting out paging and two-way radios for safety, fire, and emergency response personnel. In the 1984 interview featured in *Inc.* magazine, writer, Ellen Wojahn stated: "Carter is retired, enjoys living 'where you know everybody and his dog.' He is a commissioner of the local hospital district, chairs the Industrial Committee for Creek Lake Chamber of Commerce, and serves something as an honorary communication director for the county."

Neighbor, Roger Barker, remembers Carter as friendly and one who liked to entertain a lot, amusing you with stories of his past. Barker added that Carter, "didn't believe that the cell phone industry would work due to the short range [of frequency]," and had told him, he "was broke by the Bell System," adding, "he had respiratory problems and puffed up at the end."

On February 26, 1991, Carter, age sixty-seven, died in Presbyterian Hospital in Dallas of lung disease and was buried in Oaklawn Cemetery. Carter's legal battles contributed to the eventual breakup of AT&T and spawned a multi-billion dollar industry. Carter was known to be stubborn and would not give up on something he believed in. He believed his cause was right. Even after his settlement with AT&T, he stayed with it, often dipping into his own pockets to lobby against state or federal initiatives that might hinder the growth of the Interconnect industry. Today, all the former Bell System Operating Companies have grown larger and invested in other technologies, enriching our lives. Since the breakup of AT&T, we have experienced the internet, Bluetooth short range wireless connectivity, cellular networks providing voice and high speed data, voice-to-text, machine-to-machine communication, and the dreaded "robo-dialers." At least we are benefiting from *most* of these innovations. If it weren't for Tom Carter's stubbornness, the technology revolution may have been delayed. Former FCC Commissioner, Nicholas Johnson, who wrote the *Carterfone* opinion, states that when Carter's wife was asked what should go on his tombstone, she replied simply, "Here lays a stubborn Texan." Had Carter not been as stubborn, we might still be talking with a corded handset connected to a desk or wall phone. Of course, provided by Ma Bell. Thanks, Tom, for the doors of opportunity you opened for all of us.

Since the late 1960s several of the original Phone Warriors continue to meet socially. Many of us mentioned in the book have been married to the same women for 50 years or more. Recognition should be given to our wives for their continual support. Your indulgence in the pursuit of our goals has not gone without our utmost appreciation.

Craig Dorsey (L), the author, and Susie.

Craig Dorsey (L) and Jim Healy (R)

Bud Toly (R), telling Jerry Burns (L) the secret of selling phone systems

Left to right: Darrel Nelsen, the author, Jerry Burns and his twin, Jack

End Notes

Chapter 1

End of the Line, Leslie Cauley, pgs. 28,29. *The Fall of the Bell System*, Peter Temin, pg. 33. *The Fall of the Bell System*, Peter Temin, pgs. 29-33, *Heritage and Destiny*, Alvin von Auw, Part One, *Life in the Bell System*, Don Lively's Essay (U.S. History), *Disconnecting Parties*, W. Brooke Tunstall, pg. 7.

Chapter 2

Nicholas Johnson, "Carterfone: My Story," *Santa Clara High Technology Law Journal*, 2009, Vol. 25, Issue 3, pgs. 685 and 686. Telephone interview 7/26/18 and emails exchanged between 10/20/18-10/21/18 with Wallace Hammond, Carter's son-in-law. "Phone Company Upheld in Ban on Hush-A-Phone," *The New York Times*, Feb. 17, 1951, p. 29. Hush-A-Phone Corp. v United States, 238 F.2d 266, 269 (D.C. Cir. 1956 and see U.S. Court of Appeals for the Fifth Circuit-365 F.2d 486, August 17 1966. *Carterfone Changes Our World*, Free Online Library, Communications News, 9/1/1984. "Thomas F. Carter of Carter Electronic: Calling For Competition," Ellen Wojan, *Inc. Magazine* 4/1/84.

Chapter 3-4

The author's personal experience working with Dorsey and numerous conversations and telephone interviews 7/31/18-6/20/20.

Chapter 5

Inc. Magazine 4/1/84. Law.com.justia, U.S. Court of Appeals for the Fifth Circuit-365 F.2d 486, 8/ 17/1966.

Chapter 6

Phone interviews with Bud Toly, Craig Dorsey, Darrel Nelson during 2018 to 2020. Interview with Pat Howard 4/25/19.

Chapter 7

The Fall of the Bell System, Peter Temin, pg. 47. The Deal of the Century—The Breakup of AT&T, Steve Coll, pg. 17-119.

Chapter 8

Phone interview with Jerry Burns 8/9/18. Phone interview with Roger Williams 8/20/18. Phone interview with Jack Burns 8/21/19.

Chapter 9

Phone interviews with Wallace Hammond, Dallas Jerrell, Jason Horrell, and Peter Campbell 8/10/18 to 1/10/19. "The Secret at UCS: Managerial Acumen," *Inter-Connection*, November 1974, pg. 6.

Chapter 10

Phone interviews and email emails exchanged from 5/23/19 to 12/23/19 with Dave Perdue. Perdue also supplied copies of articles about his companies.

Chapter 11

ITT History, itt.com/About History, 2/8/18, *International Telephone Telegraph Corp.*, Encyclopedia.com. "Harold Geneen, Dies at 87," Obituary, *The New York Times*, 11/23/97. *The ITT System 12*, 100 Years of Telephone Switching.

Chapter 12

Phone interviews with Craig Dorsey, Bob Knutson, Jim Healy, Gil Engles, Steven Sherman and various Executone dealers from 8/1/18 to 12/5/19.

Chapter 13

U.S. Court of Appeals for the Second Circuit-700 F. 2d 785 Decided 1/2/83 pgs. 7, 10, 12, 13, 16, 22, 30, 31, 40.

Chapter 14

Heritage & Destiny, Alvin von Auw, pgs. 46-47, 121, 123, 152-155. See Appendix B, *The Fall of the Bell System*, Peter Temin, pg. 96 ("Alvin von Auw, drafted the speech."). *The Deal of the Century*, Steve Coll, pgs. 40-43.

Chapter 16

Telephone interviews, texts, and emails with Kelly, Engles, Sherman 5/18/18-10/29/19.

Chapter 17

Starting Up Silicon Valley, Katherine Maxfield, pgs. 3-11, 18-24, 40, 50. "Silicon Valley Success Story," *The New York Times,* 6/4/79. "Its Past Success May Haunt Coradian," *The New York Times,* 8/26/81.

Chapter 18

Against the Odds INTER-TEL, the First 30 Years, Jeffrey L. Rodengen, SEC 10 K filings, phone interviews, texts, and in-person interviews with Steve Mihaylo, Steven Sherman, Bob Knutson, Craig Dorsey 6/12/17-4/26/19.

Chapter 19

How to Turn $4,000 into Billions: Hard Work, Passion, Waterloo Regional Record, Rose Simon, 5/14/2015. "Terry Matthews: A well-connected Celt," *The Guardian,* David Gow, 10/27/2000. "There's a revolution going on, says Terry Matthews," *Ottawa Business Journal,* Sir Terry Matthews, 6/25/15. profitguide.com 6/2/2010-6/25/15. *Knights of the New Technology,* David Thomas. "Mitel," "Terry Matthews," Wikipedia.com, *Michael Cowpland,* Maclean's, John Schofield.

Chapter 20

"SD County Phone Plan Put on Hold, and Complex County Phone Plan Under Scrutiny," *San Diego Union Tribune,* Carol Sottili, 12/12/82. Ibid "$33 Million Phone System Pact is Canceled by County," 1/13/83, A-1. Ibid "Telink to Ask Revival of Phone Pact," 3/15/1983, B-1. Ibid, "Grand Jury Probes Possible Fraud in County Phone Plan," 3/22/1983 B-1. Ibid, "Supervisors Act to Revive Proposal for Private County Phone System," 8/10/83, B1. Ibid, "Phone Bribery Probe Deepens," 9/1/83, B-1. Ibid, "Telink Contract Under Probe," 9/29/83, B-1. Ibid, "Defendant Pleads Guilty in Telink Scandal-Bell Implicates ex-County Officials in Telecommunications Kickback Scheme," 5/22/1984, Wm. Polk, B-1. Ibid, "Thomas H. Bell given 3 Years in the Telink Case," 8/7/84, Bill Ott, B-3. Ibid, "Telink Scandal Figure Sentenced to Prison," 8/7/84, Wm. Polk, B-3. Ibid, "Phone Indictments Expected," 10/31/84, Bill Ott, B-1. Ibid, "Ex-Officials Indicted in County Phone Deal," 10/31/84, Wm. Polk and Jay Johnson, A-1. Ibid, "13 Indicted in Phone Fraud Case—2 Companies also Charged—Plot to Rig Bids Alleged," 11/1/84, Carol Sottili and Rivian Taylor, A-1. Ibid, "The County Phone Scandal: A Web of Bribes, Drugs, Sex," 11/1/84, Wm. Polk and Jay Johnson, A-1. Ibid, "Indicted firm May Bid on Job," 11/2/84, Carol Sottili, B-1. Ibid, "Ex-Telink Official Pleads Guilty in County Phone Probe," 11/6/84, Bill Ott, B-3.

U.S. District Court Southern District of California (records at The National Archives at Riverside), 8 CV1248, 84CR0985, 83036X, 56X.

Telephone and in-person interviews with: Craig Dorsey, Jim Healy, Jack Hart, Bob Knutson, Peter Campbell, Wallace Hammond 9/20/2018-8/10/2019.

Chapter 21

SEC annual 10K filings, telephone interview and emails with Gil Engles, Tom Kelly, Steve Mihaylo, Steven Sherman, 2/10/19-10/29/19. "ABI agrees to be acquired by TIE," *LA Times*, 12/12/87. "TIE vs. NYNEX," CaseText.com, "Marmon Group moves TIE from Seymour CT. to Overland Park," *Kansas City Business Journal*, 2/28/98.

Chapter 22

Telephone interviews and email with Alan Kessman, Wm. (Bill) Jacobson, Pete Cunnigham, 12/28/18-2/25/2020, "Tallying phone calls a cheaper way," *Business-Week*, 9/8/75.

Chapter 23

Telephone interviews, texts, and email with Alan Kessman, Steven Sherman, Kent Burgess, Craig Dorsey, Bob Knutson, Nick Visser, 2/17/19-5/17/20. Company History.com, "Firm ties to merge to success," *The Arizona Republic*, 10/10/88, "Iso(e)tec-Vodavi-Executone Marriage Bears Hope, Risk," *Intercorp*, 5/27/88, pg. 40, *The New York Times,* 4/11/96, Various SEC 10K filings.

Chapter 24

U.S. Court of Appeals for the Second Circuit-700 F. 2d 785 Decided 1/2/83.

Chapter 25

Starting Up Silicon Valley, Katherine Maxwell, pgs. 15-24, 68-71, 110—114, 178-181, 263 "Why ROLM's Star Isn't Burning as Bright," *BusinessWeek,* 8/2/82. "What's News," *Wall Street Journal,* 11/18/83. *IBM Archives: 1984.* Various SEC annual 10K filings.

Chapter 26

Telephone interviews, texts, and email with Dave Perdue, 5/23/19-12/24/19, "ATS Telephone Systems—The ROLM Mid-South distributor," *BestSellers*, Dec., Jan., Feb. 1986, published by ROLM Corp.

Chapter 27

Telephone interviews, texts, and email with Sen. Joe Bruno, and Bob Schwartz, 2/8/20-3/21/20, SEC annual Coradian10K filings. "Coradian," *Inc. Magazine* 11/1/81." The Unhappy Marriage Of Coradian And Rolm," More from *Inc. Magazine*, 11/1/81, David C. Allon. "Wescom," Wikipedia, last edited 10/3/19.

Chapter28

"Mass said in Charleston for Jarvis Corp. Founder," *Richmond Times-Dispatch*, 10/21/85. Telephone interviews and email with Christy Jarvis, Emmett Jarvis, Nick Visser, Jim Bloom, 3/31/20-5/22/20.

Chapter 29

Telephone interviews and email with Baszucki 9/25/19-1/25/20, SEC.gov/Archives/edgar, annual Norstan 10K filings, "Norstan Reports Earnings to April 30," *The New York Times*, 6/21/94.

Chapter 30

Telephone and in-person interviews and email with Mihaylo, SEC annual Inter-Tel 10K filings, *Against The Odds: INTER-TEL, the First 30 Years*, Jeffrey L. Rodengen, *Executone of Columbus, Inc. v. Inter-Tel, Inc.*, CourtListener.com, 665F.Supp.2d899, *About Steven G. Mihaylo*, eCommerce Solutions, *USA v. Inter-Tel Technologies, Inc.*, U.S. District Court, CA Northern District, Case #: 3:04-cr-00399-CRB-1, "Naming Rights and Wrongs Donor's Troubles Put CSUF on Spot," *The Orange County Register*, 2/25/07, John Gittelsohn, Marla Jo Fisher and Brian Joseph.

Chapter 31

Telephone interviews, email, and texts with Steven Sherman, Glenn Fitchet, Michael Stewart, Kent Burgess 2/7/19 -5/14/20, SEC Vodavi annual 10 K filings, "Firm Tries to Merge to Success," Newspapers.com: *The Arizona Republic*, 10/10/88, "Stan Blau Becomes President and CEO of Vodavi," Ibid, 4/16/87, "Vodavi Signs Pact to Buy Executone From Contel Business Systems," Ibid, 12/17/87, pgs. 39,81,46, "Sherman Resigns from Vodavi Board," Ibid, 1/12/98.

Chapter 32

Telephone interview and email with Alan Kessman, 2/4/20. SEC ISOETEC annual 10 K filings, "Interview with Alan Kessman," *The Arizona Republic* 7/7/88, pgs. 8, 29. "Executone Lays Off 200, Closes Norcross Headquarters," *Atlantic Constitution*, 3/2/88.

Chapter 33

Telephone interview with Dr. Ben Eisner, 3/15/20 and emails of 6/6/20-6/10/20.

Chapter 34

Telephone interviews with Pinchas Just, 310/20 and Danny Simons 6/9/230, 6/11/20.

Chapter 36

Telephone interviews with Wallace Hammond and email 7/26/18-10/22/18. *Thomas F. Carter v. American Telephone & Telegraph Co. et al,* U.S. Court of Appeals for the Fifth Circuit, 8/17/66. "Thomas F. Carter of Carter Electronics: Calling for Competition," *Inc.com,* Ellen Wojahn 4/1/84, *Carterfone Changes Our World,* The Free Library, Communication News, 9/1/84, "The Man Who Beat AT&T," *The New York Times,* "Carterfone: My Story," *Santa Clara High Technology Law Journal,* Vol. 25, Issue 3, 2009, Nicholas Johnson. Phone interviews with Lloyd Huffman, Clayton Gautreaux, Damon Steven, Roger Barker 6/14/18-6/21/18.

• ● •

Glossary

As not all readers may be familiar with some of the labels contained herein, a brief description of telephony terms and phrases used during the period this book covers might be helpful. Several terms below are also further defined in the book. The terms are listed in alphabetical order.

AT&T

American Telephone and Telegraph. Often referred to as the "Bell System" which comprised Long Lines (long-distance calling), Bell Labs (technical development), Western Electric (manufacturing), and Bell Operating Companies (providing local lines to homes and businesses and the equipment used and billing for same). Often referred to as Regional Bell Operating Companies (RBOCs) of which there were twenty-six.

Baby Bells

With the break-up of AT&T, the Bell Operating Companies became "Baby Bells" which were independent and able to merge, make acquisitions, and pursue other lines of business.

Configurations and Ports

The early KTSs and PBXs usually had a specific limitation of trunks and stations (before more equipment was required). A KTS system might have a configuration of ten by twenty-four, indicating maximum business lines and stations or multi button telephones. Larger configurations were offered by PBXs. Microprocessor-controlled systems were frequently listed in terms of "ports" that could be used for either trunks or stations in any combination not to exceed the system's number of ports.

"Cut-over"

The term refers to the moment a new telephone system in installed and the old system is disconnected and removed.

Federal Communications Commission (FCC)

The FCC was established by the Federal Government to regulate telephone utilities and issue licenses in radio frequency spectrum bands for broadcasters, emergency services, and cellular companies.

"Hardwired"

The term refers to when the new system is directly connected the operating company lines or trunks, bypassing the PCAs—also called a "direct connection."

Hook Switch and Cradle

All single line telephones (not including portable phones) had a cradle in which the handset was placed. In early rotary dial phones, lifting the handset from the cradle-initiated dial tone from the central office. Returning it to the cradle disconnected the call. The hook switch referred to two buttons in the cradle on later telephones that accomplished the same. As additional features became available in PBXs, the hook switch, when momentarily depressed created a signal to the PBX that another feature such as a "hold," "transfer," or "conference" was desired. Most phones today use a "flash" key for this function.

Hybrid Systems

As technology advanced, both KTS and PBX systems became controlled by minicomputers and microprocessors. With intelligent phones, fewer wires were needed between the central unit and the telephone station. Large twenty-five- or fifty-pair cable was no longer required. The added intelligence in the central unit meant that they could be configured as either a KTS system or a PBX with the advantages of both. Eventually, most hybrid systems also supported single-line telephones to lower overall costs. Distinctions between a KTS or PBX no longer applied. A significant advantage of the hybrid systems was that the distributor could provide additional savings by connecting business lines rather than "trunks" which were more expensive, assuming the local telephone company offered both.

Interactive Voice Response (IVR)

The term refers to a software feature or a separate system that responds to voice commands or DTMF tones to indicate the caller's intent and direct the call to the appropriate person or group. A prompt such as, "Say: claims, billing, or new quote," routes the caller to the desired extension. Many systems respond to both voice and DTMF (Touch-Tone™ signals).

Key Telephone System (KTS)

Designed for small businesses or homes with more than one line, they were controlled by a Key Service Unit (KSU) that was wall-mounted in an equipment room. The business lines appeared under buttons or "keys" on each telephone. The Western Electric telephones were offered in six-, twelve-, or eighteen-button telephone sets. One red button was dedicated as a "hold" key and another dedicated to the intercom line. With an incoming call, the line associated with a specific key would flash. When answered and the incoming caller put on "hold," the key would flash at a more rapid rate. The attendant or person assigned to answer the call would then depress the intercom key, dial the person to whom the caller wished to speak, and inform the internal party which line to depress. In the late 1970s, ten- and twenty-button telephones were also offered.

The primary advantage of a KTS system was flexibility. Any and all telephone stations could answer incoming calls. To avoid confusion, bells (ringers) were placed on the phones designated to answer incoming calls. KTS systems were much less costly than PBX systems that could handle more stations. Disadvantages were numerous. Every business line needed a dedicated key that accessed a pair of wires to each phone. Large twenty-five- or fifty-pair cables were needed to connect each multi-button telephone. This meant more labor, large conduits, or unsightly wires running through an office. All keys could be accessed by anyone at any time. The result was people picking up the wrong line and, if not put back on "hold," the caller would be disconnected. The number of phones that could be supported was also limited to the intercom which had a capacity of forty stations. Bell marketing reps were encouraged to propose larger PBX system to accommodate a subscriber's growth as they began to reach thirty stations.

Lines and Trunks

Both primarily refer to a pair of wires from the Bell Operating Company's central office to homes or businesses. The terms are often interchanged. Normally "Business Lines" were individual pairs that would terminate in small Key Systems. These groups of lines were designated sequential numbers that would "hunt" to the next line if the first was busy. The term "Line" is also used as "Station Lines," behind a PBX. With some RBOCs, the subscriber had a choice of a "measured rate line" or a "flat rate line." With a measured line in, the subscriber paid for each call made—local or long-distance. For a home subscriber with limited calling, it was usually less expensive. It was a popular choice with attorneys, as the monthly bill itemized all calls enabled them to bill back calls to his/her client. Flat rate lines were more expensive per month, but provided unlimited local calling.

Trunks were also pairs of wires from the central office, but with a slightly different polarity. Trunks more often terminated in a switching system such as a PBX. Up until the '90s, RBOCs charged more per month for trunks as compared to a business line. Rationale for this is explained in the "Background" chapter.

Private Branch Exchange (PBX)

Also see Private Automatic Branch Exchange. Located at the subscriber's place of business, all trunks terminated at the "switchboard(s)" or "attendants' position(s)." A manual PBX required a switchboard operator to answer and distribute all incoming and outgoing calls. This was accomplished by several sets of cords. An incoming call would flash at the switchboard, alerting the attendant/operator to insert one of the cords into the trunk position to answer the call. After the caller identified the person they wished to speak with, the operator would insert the other part of the cord and ring the internal party, creating a connection between the two. Calls could also be screened in this process. An internal party wishing to make an outside call would lift their hand set so the operator would connect them to an outside trunk—assuming they spotted the flashing station light on the switchboard. The operator could listen to any call in progress at any time. Hardly a "private" system. To avoid having to go through an operator to make a call, executives would request a private outside line installed on a six-button key telephone providing them with privacy. With this configuration, a KTS system would be installed behind the PBX, at an additional cost. The advantage of the PBX was that it could more readily handle growth. Another advantage was that most employees could get by with a single-line telephone, the monthly rental for which was much less that a multi-button telephone.

Private Automatic Branch Exchange (PABX)

This was a major improvement over the manual switchboard. The attendant/operator position appeared similar to the older manual cord board. More modern and sleek consoles became available later and eliminated the unsightly cords. The advantages of the automatic PBXs enabled internal parties to dial each other directly (station-to-station). To access an outside trunk, one had only to dial "9"—assuming one was available. Tie lines that connected one PBX system to another provided station-to-station dialing through a dedicated circuit.

The disadvantages were also numerous. A large switch had to be installed in a separate room. When first introduced, these were heavy mechanical switches. The early "Strowger" switches stood in racks ten feet high. Developed at the turn of the century, they were still in use in Central Offices and businesses in the 1960s and '70s. Later-developed crossbar and cross-reed switches were also large but a little less noisy. They all required their own space in the customer/subscriber's offices. They were expensive. Not only was an installation fee charged to the customer, but "basic termination fees" calculated by the local telephone company were also imposed. The termination fees were amortized over ten years. If a customer moved or outgrew their system short of the ten-year period, they would be required to pay off the balance of the fee, much like a mortgage.

The abbreviation of PABX became shortened to PBX as the manual systems were replaced.

Protective Coupler Arrangements (PCAs)

In the late 1960s, the FCC provided the approval for subscribers to use any type of telephone system or device to be connected to the AT&T network so long as it did not injure the network. AT&T argued for and received approval from the FCC to provide some type of protective device that would stop any harmful voltage or sound being transmitted. PCAs were not always provided and often created more problems. The requirement of the use of PCAs was abandoned in the early '80s

Public Utility Commissions (PUC)

State PUCs review and approve or deny requests for rate increases or additional services proposed from various utilities including telephone companies.

Subscriber

The Bell System referred to their customers as subscribers, as they subscribed monthly to their service. In the book, customer and subscriber are synonyms.

Single-Line Telephone

This is a basic telephone—the type most often found in homes with little or no features. Up until the early 1970s, all were "rotary dial." Touch-Tone™ (developed by Bell Labs in 1963) or push-button dialing telephones were not introduced until the mid and latter part of the '70s. Up to then, the Bell Central Offices could only receive "rotary pulses" and had to be modified to receive the Dual Tone Multi-frequency (DTMF) tones.

Stations

This refers to the number of telephones supported by a particular system, such as a PBX or KTS system.

Voice over Internet Protocol (VoIP)

VoIP software allowed digitized voice to be transmitted over the internet. It started gaining popularity in the late 1980s and early 1990s. Early versions had some problems, but the savings over long-distance circuit-switched calls provided primarily by AT&T were so significant that a lot of developers continued to improve the service. Today most all long-distance calls are made over the internet.

Acknowledgments

I started the effort of writing about the telecom industry when my sons suggested I write about what my business life was like when I was their age. My wife, Susie, also encouraged me to take on the task, as it might be meaningful to our grandchildren at some date. Never having written anything longer than a business plan, I naively thought it would take maybe a month. It has taken over three years to complete. There were several other manufacturers and distributors that are not mentioned in the book. Too much time has passed that locating contacts and reviving details became beyond my capability.

Few writers can write nonfiction without the help of several resources, especially contacts made over the years. My first surprise in starting the book was learning that everything I thought I knew and read about Tom Carter's background was wrong. Thanks to Wallace Hammond, Carter's son-in-law, who indulged my numerous questions, Carter's background is now more accurately stated. Thanks also to Carter's friends and professional associates of Gun Barrel, Texas. They took the time to provide colorful detail about the man whose spark (and spunk) started a revolution in the telecom industry.

The staff in the library at University of California at San Diego (UCSD) who walked me through the process of researching various news archives should also be thanked. A wonderful facility, and the staff were genuinely courteous and helpful.

The research included reading what had been previously published about the Interconnect industry. Several books have been published about the break-up of the Bell System and the growth of the "Baby Bells" that followed. Only one stood out, which described much of the back-room dealings of AT&T management and the fight that one company, MCI, had in competing in the monopoly. It was Steve Coll's fine book, *The Deal of the Century*, that provided the final incentive to write about the segment of the industry that had not been covered. It's a good read for anyone.

Most of the following I have known personally or had met. After talking with several, I wish we could have talked in person. Many of the following unselfishly gave me their time: John Barbour; Paul Baszucki; Jim Bloom; Senator Joe Bruno (and daughter Kate, and Executive Assistant, Nicole); Jerry Burns and twin brother, Jack; Kent Burgess; Peter Campbell; Pete Cunningham; Craig Dorsey; John Eddy; Ben Eisner; Gil Engles; Glenn Fitchet; Ray Glynn; Jack Hart; Jim Healy; Pat Howard; Jason Horell; Dr. Ben Eisner; Wm. (Bill) Jacobson; Dallas Jarrel; Cristy Jarvis; Emmett Jarvis; Pinchas Just; Alan Kessman; Tom Kelly; Bob Knutson; Steve Mihaylo; Darrel Nelsen; Earl Nelson; Dave Perdue; Robert (Bob) Schwartz; Steven Sherman; Dannie Simons; Mike Stewart; and Nick Visser.

It should be noted that several individuals, mostly mentioned above, have passed on during the writing of the book: Helen Carter, Tom's wife; Stanley Blau, a catalyst in the industry who helped start our national association, NATA, an integral part in several companies, and a friend to all who knew him; also, Richard Long, the dynamic, lanky Texan from Fisk Electric who led NATA for several years. Last year we lost both Gil Engles, the TIE VP who was always a gentleman, and Joe Bruno, the NY State Senator and VP of Coradian.

Thanks to my wife, Susie, who made several suggestions and caught a multitude of errors ... and the quality staff at Acorn Publishing without whose help this might not have come to fruition ... and any reader who got this far in the book.

Made in the USA
Las Vegas, NV
27 July 2021

27113980R00194